Laboratory Manual

to accompany

CONCEPTUAL

Physical Science

SECOND EDITION

Paul G. Hewitt

John Suchocki

Leslie A. Hewitt

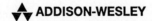 ADDISON-WESLEY

Menlo Park, California • Reading, Massachusetts
New York • Harlow, England • Don Mills, Ontario
Amsterdam • Madrid • Sydney • Mexico City

Sponsoring Editor: Sami Iwata
Publishing Assistant: Bridget Biscotti-Bradley
Production Supervisor: Larry Olsen
Cover Designer: Yvo Riezebos

ISBN 0-321-03537-2

2 3 4 5 6 7 8 9 10—CRS—03 02 01 00 99

Addison Wesley Longman
2725 Sand Hill Road
Menlo Park, California 94025

Introduction

The value of laboratory experiments in science is well established. The value of activities, however, is not so well established. If it's true that we retain something like 5% of what we read, and about 75% of what we do, then the value of activities such as those in this manual is obvious. Actually dunking an object in water and noting both the displacement and the "loss of weight" is an important experience for a student beginning the study of Archimedes' principle. Building molecular models is an important experience for a student learning how atoms bond. Likewise seeing crystals form in a microscope is important for one beginning to study minerals, and tracing ellipses with string greatly benefits one beginning the study of planetary motion. The activities in this manual set the stage for comprehending concepts treated in the textbook. Optimal learning occurs when the order of instruction is (1) activities, (2) textbook, and (3) experiments. This is in accord with the learning cycle popular in high schools, pioneered by physics educator Robert Karplus of the University of California at Berkeley a quarter century ago.

Acknowledgments

Many of the physics labs and activities in this manual originated with Paul (Pablo) Robinson, author of the lab manuals for both the high school and college versions of *Conceptual Physics*. So we owe a big thank you to Pablo.

Thanks to Earl R. Feltyberger, of Nicolet High School in Glendale, WI for the activity *By Impulse*.

For Part 6 labs we thank Carl D. Baer of Providence College for *Crystal Clear*, and Ted Brattstrom for *Mystery Powders*. For *Sugar and Sand* we are grateful to Erwin W. Richter.

For contributions to nearly all the Part 7 labs we are grateful to Leslie's husband, Bob Abrams, also a geologist. For valuable feedback and suggestions we thank CCSF physics and geology instructor Jim Court. For the simple technique of measuring specific gravity with electronic balances in the activity *Specific Gravity*, we acknowledge the note by Paul L. Willems in The Physics Teacher magazine (vol. 36, Jan. 88). For other geology resources we are grateful to AGI/NAGT (American Geological Institute and the National Association of Geology Teachers).

For Part 8 we are thankful to Ted Brattstrom for *Reckoning Latitude*, and Forest Luke for *Tracking Mars*. Both Ted and Forrest are high school teachers and amateur astronomers in Hawaii — Ted at Pearl City High School, and Forrest at Leilehua High School. For valuable feedback and advice on most of the astronomy material we are grateful to Richard Crowe, University of Hawaii at Hilo, and to John Hubisz, North Carolina State University.

For general suggestions on all of this manual we remain indebted to mentor Charlie Spiegel. For feedback we are grateful to Marshall Ellenstein.

Table of Contents

Introduction

Part 1 Mechanics

Part 2 Heat

Part 3 Electricity and Magnetism

Part 4 Sound and Light

Part 5 The Atom

Part 6 Chemistry

Part 7 Earth Science

Part 8 Astronomy

Appendix

CONCEPTUAL **Physical Science**

[**Activity**]

Introduction to Physical Science

Tuning the Senses

Purpose

To tune your senses of sight and sound.

Scientists' original source of information about the universe comes from personal observations. This leads to questioning reasons and causes. A scientist notices something, asks questions, and then tries to answer them. By this definition, can't we all be scientists?

Required Equipment and Supplies

Notebook, pen or pencil, candle, matches, and patience.

Discussion

Galileo wrote, "In questions of science the authority of a thousand is not worth more than the humble observation and reasoning of a singel individual." We'll do two simple activities; the first to tune our hearing, and the second to tune seeing.

Procedure

Perform the following two activities on your own and then answer the questions on the following page.

Activity 1: Audition of the Environment

Go outside and find a comfortable place to sit. Use only your sense of hearing, and listen with eyes closed most of the time, for about 10 minutes. Write down the sounds you hear.

Activity 2: Observation of a Burning Candle

Remain absolutely quiet while observing an unlit candle for two minutes. Record your observations.

Light the candle, observe it for two minutes, then record your observations.

After you have thoroughly exhausted all observations possible, extinguish the candle.

Observe the extinguished candle while recording more observations.

What other activities can you invent to tune your senses?

Be creative and tune your senses every day!

Questions for Activity 1: Audition of the Environment

1. What was the quietest sound you heard?

_____

2. Are there any continuous sounds *right now* (like that of an air conditioner) that you have simply tuned out? What are they?

_____

3. Put your hands over your ears *right now*. What was the last sound you heard?

_____

4. How many sounds heard did you describe?

_____

5. What sounds contained both high and low pitches?

_____

6. Did you idnetify sounds by their possible source or did you spell them out phonetically?

_____

7. How might you describe a sound to some one who was unfamiliar with its source (For example, the sound of a car to some one who had never heard of a car before)?

_____

Questions for Activity 2: Observation of a burning candle

8. Did you detect any odors?

_____

9. What color was the molten wax?

_____

10. Did you note the time and date of your candle observations.

_____

11. Compare the relative hotness 1 inch above the candle flame, and 1 inch to the side.

_____

12. Did the candle make any sounds? What were they?

_____

13. What were the patterns of the smoke before and after the flame was extinguished?

_____

14. Where was the wick actually burning?

_____

15. Was the color transition of blue to yellow in the flame gradual or abrupt?

_____

16. Did you describe the candle by drawing a picture of it?

_____

CONCEPTUAL **Physical Science** | **Activity**

● *Scientific Method*

Making Cents

Purpose
To investigate the effect of aging on a penny relative to its mass.

Required Equipment and Supplies
10 pennies per student
balances
graph paper

Discussion
The scientific method is an effective way of gaining, organizing, and applying new knowledge. The method is essentially as follows:

1. Recognize a problem.
2. Make an educated guess—a hypothesis.
3. Predict the consequences of the hypothesis.
4. Perform experiments to test predictions. If necessary, modify the hypothesis in light of experimental results. Perform more experiments.
5. Formulate the simplest general rule that organizes the three main ingredients— hypothesis, prediction, experimental outcome.

Procedure
Step 1: Propose a hypothesis to the following question (problem): What effect does time have on a penny's mass?

Step 2: Based upon your hypothesis, predict the general form of a graph that plots the mass of a penny (*y*-coordinate) relative to its age (*x*-coordinate). Make no measurements before you predict your graph.

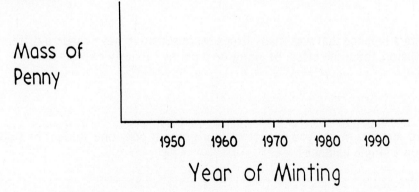

Step 3: Using a balance, measure the mass of at least 10 individual pennies minted in different years. Enter the mass in grams of each penny relative to the year it was minted (Table 1).

Table 1.

Mass of Penny:									
Year Minted:									

Step 4: Pool your data with that of all other students on the class chalkboard and create your own graph using all this data. Show the mass of each penny in grams on the y-coordinate and the year the penny was minted on the x-coordinate. Alternatively, data may be entered into a computer program that will plot the graph for you.

Step 5: Are the experimental results consistant with your hypothesis? If not, propose a new hypothesis.

Step 6: If you have formed a new hypothesis, what additional measurements might you take to support this new hypothesis? Perform these measurements and record your results and observations here:

Summing Up

1. What conclusion can you draw from the results of your experimental data?

2. What effect might aging have on the mass of a nickel, a dime, a quarter?

3. Would using a balance that was many times more sensitive have made a difference in your conclusion about the effect of aging on a penny? Briefly explain.

4. What improvements might you expect in your graph if only one student had done all the weighing on a single balance?

Making Cents

CONCEPTUAL **Physical Science** | Activity |

Accelerated Motion

Reaction Time

Purpose
To measure your personal reaction time.

Required Equipment and Supplies
dollar bill
centimeter ruler

Discussion
Reaction time is the time interval between receiving a signal and acting on it — for example, the time between a tap on the knee and the resulting jerk of the leg. Reaction time often affects the making of measurements, as for example in using a stopwatch to measure the time for a 100-m dash. The watch is started after the gun sounds and is stopped after the tape is broken. Both actions involve reaction time.

Procedure
Step 1: Hold a dollar bill so that the midpoint hangs between your partner's fingers. Challenge your partner to catch it by snapping his or her fingers shut when you release it.

The distance the bill will fall is found using

$$d = \tfrac{1}{2}at^2$$

Simple rearrangement gives the time of fall in seconds

$$t^2 = \frac{2d}{g}$$

$$t = \sqrt{\tfrac{2}{980}}\sqrt{d}$$

$$t = 0.045\sqrt{d}$$

(for *d* in cm, *t* in s, we use $g = 980 \text{ cm/s}^2$)

Step 2: Have your partner similarly drop a centimeter ruler between your fingers. Catch it and note the number of centimeters that passed during your reaction time. Then calculate your reaction time using the formula

$$t = 0.045\sqrt{d}$$

where *d* is the distance in centimeters.

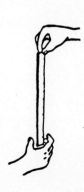

Reaction time =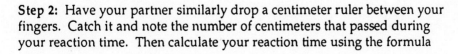

Summing Up

1. What is your evidence for believing or disbelieving that your reaction time is always the same? Is your reaction time different for different stimuli?

_____

_____

2. Suggest possible explanations why reaction times are different for different people.

_____

_____

3. When might reaction time significantly affect measurements you might make using instruments for this course? How could you minimize its role?

_____

_____

4. What role does reaction time play in applying the brakes to a car in an emergency situation? Estimate the distance a car travels at 100 km/h during your reaction time in braking.

_____

_____

5. Give examples for which reaction time is important in sports.

_____

_____

CONCEPTUAL Physical Science ──────────────── **Experiment**

Graphical Analysis of Motion

Graphing with Sonar

Purpose
To make qualitative interpretations of motions from graphs.

Required Equipment and Supplies
sonic ranging device with appropriate equipment and software
computer
masking tape
marking pen or pencil
ring stand
heavy beach ball
dynamics cart
can of soup
board for inclined plane
pendulum clamp

Discussion
A bat flies about in darkness without bumping into things by sensing the echoes its squeaks. These squeaks reflect off walls and objects, return to the bat's head, and are processed in its brain to provide the location of nearby objects. The automatic focus on some cameras works on much the same principle. These cameras use the device used here, which measures the time that ultra high frequency sound waves take to go to and return from a target object. The data are fed to a computer where they are graphically displayed. The program can display the data in three ways: distance versus time, velocity versus time, and acceleration versus time. Imagine how Galileo and Newton would marvel at such technology!

Procedure
Your instructor will provide a computer with a sonic ranging program installed. Check to see that the sonic ranger is properly connected and operating reliably. Place the sonic ranger on a desk or table so that its beam is about chest high. (Note: Sometimes these devices do not operate reliably on top of computer monitors.)

Select the computer option that plots distance vs time. Set the graph duration to "Continuous" and the maximum range to 4 meters. The sonic ranger has a *minimum* range of about 0.4 meters. Readings less than this range will be erratic. Make small pencil marks or affix a piece of string to the floor in a straight line from the sonic ranger. Point the sonic ranger at a student standing at the 5 meter mark. Mark where the computer registers 0 m, 1 m, 2 m, etc. Set the maximum range appropriately between 4 to 6 meters.

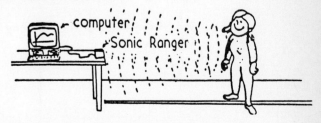

Play around and note how your motions are displayed on the screen of the computer. Note how steady motion differs from accelerated motion, and how motion toward the sonic ranger detector differs from motion away from the detector. This is truly a wonderful way to understand graphs of motion. You can make them in real time!

Part A: Analyzing Motion Plots

Step 1: Now let's analyze motion more carefully. Stand on the 1-meter mark, face the sonic ranger detector and watch the monitor. Walk away from the detector slowly and observe the real-time plot of distance vs time. Repeat, walking away from the detector faster, and observe the graph.

1. Make a sketch (or printout if a printer is available) of each graph. How do the graphs compare?

Step 2: Stand at the far end of the string. Slowly approach the sonic ranger and observe the distance-vs-time graph plotted. Repeat, walking faster, and observe the graph plotted.

2. Make a sketch of the graph for slow approach, and another for fast approach. How do the graphs compare?

Step 3: Walk away from the sonic ranger slowly; stop; then approach the sonic ranger quickly.

3. Sketch the shape of the resulting distance-time graph. What differences are evident in the two slopes?

Step 4: Repeat Step 1, but select the computer option to display a plot of velocity vs time.

4. Make a sketch of the graph. How do the distance-vs-time and velocity-vs-time graphs compare?

Step 5: Repeat Step 2, but select the option to display to plot velocity-vs-time.

5. Make a sketch of the graph. How do the two new graphs compare?

Step 6: Repeat Step 3, but select the option to display to plot velocity-vs-time.

6. Sketch the shape of the resulting graph. How do the two new graphs compare?

Part B: Motion-on-an-Incline Analysis

Step 7: Set up the sonic ranger as shown at the right so you can analyze the motion of a dynamics cart or something like a soup can that is rolled up an incline and allowed to roll back. Make sure the can is always at least 0.4 meters away from the sonic ranger. Do NOT activate the sonic ranger yet! Instead, **predict** the general shapes of distance versus time and velocity versus time graphs for this up and down motion.

soup can

Sonic Ranger

7. Make a sketch of your predicted graphs.

Select the option (if available) of both distance vs time and velocity vs time displayed simultaneously on the monitor.

8. Sketch the shape of the distance-vs-time graph and velocity-vs-time graphs.

9. Is the velocity-vs-time graph a straight line? Why or why not?

_____

_____

Graphing with Sonar

10

Part C: Analyze Pendulum Motion

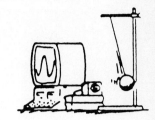

Step 8: Use the sonic ranger to analyze the motion of a pendulum as shown. Be sure the sonic ranger is always a minimum of 0.4 meters from the swinging bob. Use a pendulum with a length great enough (such as a meter or longer) to make the path of the bob nearly a straight line in the beam of the sonic ranger. Collect data of the swinging pendulum, then choose the option so that both distance vs time and velocity vs time graphs are displayed simultaneously.

10. Sketch the shape of the two graphs as they are displayed on the monitor.

11. Where is the speed greatest for the swinging pendulum bob? Least?

_____

Part D: Analyzing the Acceleration of Gravity

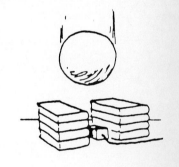

Step 9: Place the sonic ranger on the floor and stack a few books around to protect it, or inside a wire-mesh cage. Then drop large heavy objects (so that the effect of air friction is negligible) like heavy beach balls directly on it.

12. Describe and compare the distance vs time and velocity vs time graphs.

_____

_____

Summing Up
13. Did you find the acceleration of the dropping object constant (or nearly so)?

_____

14. How does this compare to the acceleration of the can rolling along the incline in Part B?

_____

_____

CONCEPTUAL **Physical Science** | Activity

Newton's Laws

Dropping and Dragging I

Purpose
To observe the effects of applied force and inertia.

Required Equipment and Supplies
about 6 feet of twine
table
3 textbooks, such as *Conceptual Physical Science* (or 3 equivalent masses)

Discussion
When a net force is applied to an object, its motion changes — it *accelerates*. Gravity between an object and the earth (the object's weight) provides a convenient virtually constant force. Drop a book and the interplay of gravity and the book's mass gives it an acceleration g (nearly 10 m/s^2). If the dropping book is attached by a length of twine to another book that lies on a horizontal surface, the acceleration of the dropping book is less — it is restrained by the extra mass of the book above, and friction between the book above and the supporting surface.

Procedure
Place a book, call it Book A, on a horizontal surface. It doesn't accelerate because gravity on it is balanced by an upward support force by the table (called the normal force). So with no net force in the vertical direction, there is no vertical acceleration. But if a length of twine is attached to another book that falls vertically, the falling book, call it Book B, provides a force that accelerates both books. So tie two books together by a 3 or 4-foot length of twine, and place Book A in the middle of a table and hold Book B slightly off the edge of the table. Let go, and the books accelerate —Book B vertically, and Book A horizontally. Because they're attached, both have the same amount of acceleration. You'll want to make the connecting twine between the books slightly longer than the height of the table, so that Book A can coast to the edge after Book B hits the floor. (If you're antsy about damaging your textbooks, you could substitute other more durable masses for this activity.)

Make an estimated judgement (guesstimate) about the amount of acceleration the two-book system undergoes compared to that of free fall, g. Then try it and see.

Significantly, there is more than the horizontal force of the string (tension) acting on Book A. That's friction, which acts in a direction opposite the motion. If friction is more than the tension, the system doesn't even move when you release Book B. Do what you can to minimize friction on Book A, like using rollers of some kind (round pencils, perhaps). If you mount Book A on a cart, be sure to attach an equal amount of added mass to Book B!

(If you use masses other than books; the main idea is to first try equal masses for A and B, and to minimize friction between A and the surface and between the string and the edge of the table.)

Going Further

1. Place another book on Book A, doubling its weight (*and doubling the friction between it and the table also!*). You now have the force provided by the weight of one book acting on three books — three times the mass. Here's where its important that you've minimized friction! Predict the fraction of *g* that the books will experience when you release Book B.

2. Repeat, but this time attach the extra book to falling Book B. Now you have twice the force acting on three times the mass (and less friction compared to the previous trial!). Predict the amount of acceleration, compared to the previous cases, and then try it and see.

Summing Up

1. Rank the acceleration for each case:

   Most acceleration _____

   Mid acceleration _____

   Least acceleration _____

2. When a book is in free fall, the only force on it is gravity (its weight), which acts on the mass of the single book — and it accelerates at *g*. If you drop two books, attached by a length of string, twice as much gravity force acts on twice the mass — and it accelerates at:

   _____

3. When one of the books is dropped while connected by twine to the other book atop a horizontal table, both books accelerate. Why is the acceleration less than *g*?

   _____

4. When two books were dropped while connected by twine to a single book atop the table, the acceleration was greater than for one book dropping. If three books were dropped while attached to a single book atop the table, would the acceleration be even greater?

   _____

5. Apparently, the greater the load that drops compared to the load dragged across the table top, the greater the acceleration. What is the upper limit of acceleration that can be produced by this method? (Use exaggeration: Suppose the dropped load was huge and the sliding mass very small.)

   _____

6. How much did friction affect your results? How could you better reduce friction?

   _____

Dropping and Dragging I

CONCEPTUAL **Physical Science** ──────────────────── **Experiment**

Newton's Laws

Dropping and Dragging II

Purpose
To measure the effects of applied force and inertia.

Required Equipment and Supplies
about 6 feet of twine
table
3 carts (or equivalent), each about the mass of the *Conceptual Physical Science* textbook
pulley
meter stick
stopwatch

Discussion
In *Dropping and Dragging I* we observed the accelerations of three different systems of books — one dropping and dragging another across a table top, one dropping and dragging two books across the table top, and finally two books dropping and dragging a single book across the table top. *Dropping and Dragging I* is a qualitative activity — estimating rather than measuring. This experiment is a repeat, only quantitatively. In place of books, you'll use carts and a pulley so that friction will not play as large a role. If friction can be made negligible, the relationship of force to mass on acceleration is more clearly seen.

Do what you can to minimize friction in this experiment. And how you measure and compare accelerations is up to you.

Procedure
Place a cart, call it Cart A, on a horizontal surface. It remains at rest because there is no force acting on it in a horizontal direction — yet. But when twine is attached to another same-mass cart that falls vertically, the falling cart, call it Cart B, provides a force that accelerates a tow-cart system. So tie two carts together by a 3- or 4-foot length of twine, and place Cart A in the middle of a table and hold Cart B slightly below the pulley as shown in the sketch. Let go, and the carts accelerate — Cart B vertically, and Cart A horizontally. Because they're attached, both have the same magnitude of acceleration. (You can use an equivalent mass for Cart B.)

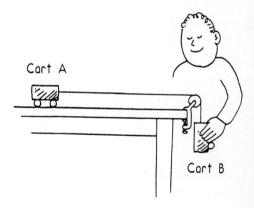

Cart A

Cart B

Estimate the acceleration of the two-cart system compared to that of free fall, g. Enter your prediction:

    Predicted acceleration of the 2-cart system _____

Suggest a way to measure the accelerations of free fall and the acceleration of the 2-cart system. Check with classmates — go to your instructor as a last resort.

    Observed and measured acceleration of the 2-cart system _____

Dropping and Dragging II

Going Further

1. Place and attach another cart atop Cart A, doubling its mass (*and doubling the friction between it and the horizontal surface also!*). You now have the force provided by the weight of one cart acting on itself and the two carts atop the table — three times the mass. Predict the fraction of *g* that the carts will experience when you release Cart B.

   Predicted acceleration of system with doubled Cart A: _____

   Measured acceleration of system with doubled Cart A:_____

2. Repeat, but this time attach the extra cart to falling Cart B. Now you have twice the force acting on three times the mass.

   Predicted acceleration of system with doubled Cart B: _____

   Measured acceleration of system with doubled Cart B:_____

Summing Up

1. For one cart in free fall, the ratio *weight/mass* is *g*. If we drop two carts (tied together) we double both the weight and the mass, and acceleration is:

   _____

2. In this experiment we doubled the mass without doubling the weight that caused acceleration. By placing one cart on the table we in effect canceled its weight by the table's support force. The only weight that caused acceleration was that of the dropping cart. In this case the ratio of *weight/twice mass* produces an acceleration of:

   _____

3. When the dragged mass is doubled, the single dropping mass pulls on three times its mass (its mass + the mass of the other two carts). If friction plays a small role, the acceleration of the system is about:

   _____

4. When the weight of two dropping carts is the force, and they are attached to one cart on the table, the force is twice the previous case, and the mass being accelerated is the same — that of 3 carts. With negligible friction, the acceleration of the system is:

   _____

5. What acceleration occurs when two falling carts are attached to two carts on the table?

   _____

6. How much did the pulley and surface friction affect your results? How could you reduce friction?

   _____

Dropping and Dragging II

CONCEPTUAL **Physical Science** ┌─────────┐
│ Activity │
└─────────┘

Impulse Equals Change in Momentum

By Impulse

Purpose

To investigate the effect that a prolonged stopping time has on force, when momentum changes.

Required Equipment and Supplies

table
meter stick
short lengths (30 to 40 cm) of "non-strong" string
various masses

Discussion

In bungee jumping, the cord that supports the jumper must stretch. If the cord doesn't stretch, then when fully extended it either brings the jumper to a sudden halt, or the cord breaks. In either case, OUCH! Whenever a falling object is brought to a halt, the force that slows the fall depends on the time it acts. We will see evidence of this in this activity.

Procedure

Through the small hole at the end of a meter stick, thread a piece of string about 30 cm long and tie it in place. (If there is no hole, consider drilling one, or tying the string very tightly.) Attach a mass to the other end of the string — try a kilogram for starters. Place the stick on a table top and slide the stick so that most of its length extends over the edge of the table, Figure A. While holding the stick firm to the table, hold the mass slightly beyond the edge of the stick — then drop the mass. Its fall should be stopped by the string and the bending of the stick. The string shouldn't break. (If it does break, try a stronger string or a smaller mass — experiment). Now repeat this activity, but with only a small portion of the stick extending over the table's edge. What happens now?

Fig. A

Fig. B

Try different configurations to see what conditions result in the string breaking. Experiment with different masses, different amounts of meter stick overhang, and different string lengths.

Summing Up

1. Why is it important that a bungee jumper be brought to a halt gradually?

_____

_____

2. How does the *impulse = change in momentum* formula, $Ft = \Delta mv$, apply to this activity?

_____

_____

3. Exactly why did the string break when there was less "give" to the meter stick?

_____

_____

4. The breaking strength of the string most certainly plays a role in this activity. How does the length of the string play a role?

_____

_____

5. How does the falling mass play a role. Will twice the mass require twice the stopping force if it is brought to a halt in the same time? Defend your answer.

_____

_____

6. Why is it important that fishing rods bend?

_____

_____

CONCEPTUAL Physical Science

Energy

Energy Ramp

Purpose
To investigate the relationship between the height of a ball rolling down an incline and its stopping distance when it rolls off the incline.

Required Equipment and Supplies
about a 2-meter ramp (5/8 inch aluminum channel is fine) with support about 1/2 meter high
steel or glass ball
meterstick
carpet floor, or piece of carpet strip (about 4 m long and about 50 cm wide)

Discussion
Mechanical energy is the product of force and distance moved by the force. When we exert a force to change the energy of an object, we do *work* on an object. A measure of that work, or change in energy, is force x distance. Elevate a ball and we do work on it. With a force equal to its weight we lift it a certain height against gravity. The work done gives it *energy of position* — gravitational potential energy (PE). Raise it twice as high and it has twice the energy. Another way of saying it has twice the energy is to say it has twice the ability to *do* work. When it rolls to the bottom of the incline it can do work on whatever it interacts with. If it has twice the energy, it can do twice the work. An easier-to-visualize example is that of a crate sliding onto a factory floor. In sliding, it does work on the floor and heat it up as it skids. This work is the force of friction x distance of sliding. The question is raised: Will it skid twice as far if it has twice the energy? We'll answer this question not by sliding crates down a ramp (for much of their PE would go into heating the ramp itself, which complicates matters), but by rolling balls down inclines and then onto a carpet.

Procedure
Step 1: Divide the ramp into 4 equal spaces, and mark them, as shown in Figure A. Assemble the ramp so that when you roll a ball down it, the length of carpet it rolls onto is sufficient to stop it. Experiment and see.

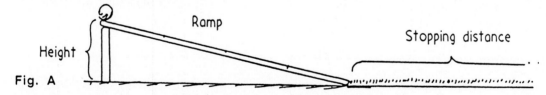

Fig. A

Step 2: Release the steel ball at each of the intervals along the ramp. Measure the vertical height from the floor or table. Roll the ball three times from each height and record the stopping distances in Data Table A.

Step 3: Change the angle of the ramp, but launch the ball from the SAME VERTICAL HEIGHT that you did for the previous ramp position.

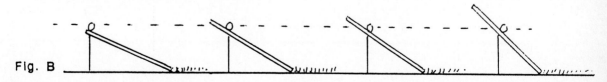

Fig. B

STEP 2 DATA

Step 2 data: Ball's initial ramp position (cm from the bottom end)	Initial vertical height of the bottom of the ball above carpet (cm)	Stopping distance (cm)			Average stopping distance (cm)
		Trial 1	Trial 2	Trial 3	
30 cm					
60 cm					
90 cm					
120 cm					

STEP 3 DATA

Step 3 data: Ball's initial ramp position (cm from the bottom end)	Initial vertical height of the bottom of the ball above carpet (cm)	Stopping distance (cm)			Average stopping distance (cm)
		Trial 1	Trial 2	Trial 3	

Going Further On the graph paper provided by your instructor, construct a graph of average stopping distance (vertical axis) versus launch height (horizontal axis) for the ball in Step 2.

Summing Up

1. Does your data indicate direct proportions between height of release and stopping distance? (That is, does twice the release height give twice the stopping distance)?

_____

2. In terms of work and energy, interpret your results from Step 3 where your release of the ball was from the SAME release height but from ramps sloped at DIFFERENT angles. Do these results make sense? Explain.

_____

3. What conclusions (if any) can you make about the SPEED of the ball at the bottom of each differently sloped ramp? Explain

_____

4. Nowhere in this activity is the mention of kinetic energy — energy of motion (KE). A little study of energy conservation tells us that the gain in KE of the ball as it rolls down the ramp is equal to the decrease in PE as the ball loses height — in short, $\Delta PE = \Delta KE$. Why were we able to bypass KE in our analysis here? (The answer to this question underlies the reason that physics types use energy principles to solve problems — intermediate steps often need not be considered!)

_____

_____

Energy Ramp

CONCEPTUAL **Physical Science** | **Experiment**

Projectile Motion

Bullseye

Purpose
To predict the landing point of a projectile.

Required Equipment and Supplies
1/2" (or larger) steel ball
empty soup can
meter stick
stopwatch
means of projecting the steel ball horizontally at a known velocity

Discussion
If you were to toss a rock in some region of gravity free outer space, it would just keep going — indefinitely. The rock would continue its motion at constant speed and cover a constant distance each second (Figure A). When motion is uniform, the equation for distance moved is

$$d = \bar{v}t$$

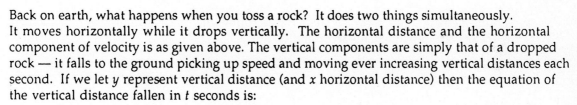

Let x be the horizontal distance, and y the vertical distance. Then speed is

$$\bar{v} = \frac{x}{t}$$

Back on earth, what happens when you toss a rock? It does two things simultaneously. It moves horizontally while it drops vertically. The horizontal distance and the horizontal component of velocity is as given above. The vertical components are simply that of a dropped rock — it falls to the ground picking up speed and moving ever increasing vertical distances each second. If we let y represent vertical distance (and x horizontal distance) then the equation of the vertical distance fallen in t seconds is:

$$y = \tfrac{1}{2}gt^2$$

where g is the acceleration due to gravity. Starting from rest, the instantaneous falling speed v after time t is

$$v = gt$$

When we toss a rock horizontally, all four equations above apply. The result is curved motion (Figure B). The vertical component of motion undergoes acceleration, while the horizontal component does not. The secret to analyzing projectile motion is to keep two separate sets of "books": one that treats the horizontal motion according to

$$x = \bar{v}t$$

and the other for vertical motion according to

$$y = \tfrac{1}{2}gt^2$$

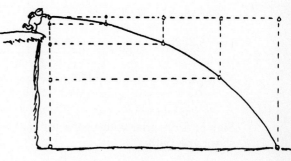

Horizontal Motion

• When thinking about how *far*, think $x = \bar{v}t$

• When thinking about how *fast*, think $\bar{v} = \dfrac{x}{t}$

Vertical Motion

• When thinking about how *far*, think $y = \frac{1}{2}gt^2$

• When thinking about how *fast*, think $v = gt$

When engineers build bridges or skyscrapers they do *not* do so by trial and error. For the sake of safety and economy, the effort must be right the *first* time. Your goal in this experiment is to predict where a steel ball will land when projected horizontally from the edge of a table. The final test of your measurements and computations will be to position an empty can so that the ball lands in the can on the *first* attempt.

Procedure

Step 1: Use a means of projecting the steel ball at a known horizontal velocity (it may be a spring gun, a ramp, or some other device). Position the ball projector at the edge of a table so the ball will land downrange on the floor. Do NOT make any practice shots (the fun of this experiment is to *predict* where the ball will land without seeing a trial!) You will be given the initial speed of the steel ball (or perhaps you'll have to devise a way to measure it). In any event, record the firing speed here.

Horizontal speed = _____ cm/s

Step 2: Carefully measure the vertical distance y the ball must drop from the bottom end of the ball projector in order to land in an empty soup can on the floor.

1. Should the height of the can be taken into account when measuring the vertical distance y? If so, make your measurements accordingly.

_____

Step 3: Using the appropriate equation from the discussion, find the time t it takes the ball to fall from its initial position to the can. Write the equation that relates y and t.

Equation for vertical distance: _____

Step 4: The range is the horizontal distance a projectile travels, x. Predict the range of the ball. Write the equation you used to predict the range. Write down your predicted range.

Equation for the range: _____

Predicted range x = _____

Now place the can on the floor where you predict it will catch the ball.

Step 5: Only after your instructor has checked your predicted range and your can placement, shoot the ball.

Bullseye

Summing Up

1. Did the ball land in the can on the first trial? If not, how many trials were required?

_____

2. What possible errors would account for the ball overshooting the target?

_____

_____

3. What possible errors would account for the ball undershooting the target?

_____

_____

Alternate Procedure

Suppose you don't know the firing speed of the steel ball. If you go ahead and fire it, and then measure its range rather than predicting it, you can work backward and calculate the ball's initial speed. This is a good way to calculate speeds in general! Do this for one or two fired balls whose initial speeds you don't know.

Summing Up

1. What relationship did you observe between firing speed and range?

_____

2. Consider the pitcher throwing the ball below. If he is conveniently on a tower so that the ball is 5 m off the ground when thrown horizontally, and the ball lands 18 m downrange, then the speed of the ball is easily calculated. What is this speed, and why does the 5-m elevation make the calculation convenient?

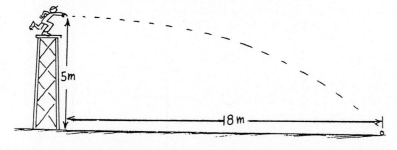

_____

3. If the projected ball is not fired horizontally, but is fired at an angle above the horizontal, the problem is more complicated. What are some of these complications, and how can they be minimized?

_____

_____

CONCEPTUAL **Physical Science**

Archimedes' Principle

Sink or Swim

Purpose
To observe the effect of density on various objects placed in water.

Required Equipment and Supplies
equally massive blocks of lead, wood, and Styrofoam
at least two aluminum cans of soda pop; both diet and regular
aquarium tank or sink
egg
wide-mouth graduated cylinder
bowl
salt
spoon
balance

Discussion
Do you know or have you observed in swimming areas that some people have difficulty floating while others don't? This activity should increase your understanding of this disparity.

Procedure
Step 1: Balance equal masses of lead and wood, using a double-beam balance. Repeat, using an equal mass of Styrofoam.

1. How do the *volumes* compare? How do the *densities* therefore compare?

_____

_____

Step 2: Try floating cans of diet and regular soda pop.

2. Which ones float? Sink?

_____

3. How does the density of the different kinds of soda pop compare to the density of tap water?

_____

_____

4. Hypothesize how the relative densities relate to sugar content.

_____

_____

Step 3: Use a balance to measure the mass of an egg. Using a wide mouth graduated cylinder, carefully determine the volume of the egg by measuring the volume of water it displaces when it is slowly (gently!) lowered into the graduated cylinder. Calculate its density: $d = m/V$.

Density = _____

Step 4: Now try to float the egg in a bowl of water. Does it float? If not, dissolve enough salt in the water until the egg floats.

5. How does the density of an egg compare to that of tap water?

_____

_____

6. To salt water?

_____

_____

Summing Up

7. Does adding salt to the water make the water less dense or does it make it more dense? How?

_____

_____

_____

_____

8. Why do some people find it difficult to float while other don't? What evidence can you cite for the notion that it is easier to float in salt water?

_____

_____

Sink or Swim

CONCEPTUAL **Physical Science** | **Activity**

Archimedes' Principle

Eureka!

Purpose
To investigate the displacement of water by immersed objects.

Required Equipment and Supplies
35-mm film canisters and string
ballast material (nuts, BBs, sand, etc.)
beam balance
500 mL graduated cylinders

Discussion
Archimedes is much remembered for his clever discovery of finding the volume of irregularly-shaped objects — like himself! He discovered this while in the public baths of Athens, pondering a problem given to him by the king — how to determine whether a particular gold crown was really 100% gold or not. His discovery is based on a very simple idea that many people don't fully understand. If perchance you are one of them, after doing this activity, you won't be!

Procedure
Your instructor will give you two film canisters, each with contents of different mass, and each with a piece of string attached. Find the mass of each canister.

Mass of lighter canister, m = _____ g

Mass of heavier canister, m = _____ g

Procedure
So that you can easily mark the water level in a graduated cylinder, attach a vertical strip of masking tape along its outer side. Fill the cylinder about three-quarters full of water. Mark the water level on the tape. Submerge the lighter canister and mark the new water level on the tape. Remove the canister.

Predict how the water level for the heavier canister will compare when it is submerged

_____

Try it and see! Was your prediction correct?

_____

Summing Up
Describe and explain your observations.

_____

_____

CONCEPTUAL **Physical Science**

Archimedes' Principle

Boat Float

Purpose
To investigate Archimedes' principle and the principle of flotation.

Required Equipment and Supplies
spring scale
triple-beam balance
string and masking tape
rock or hook mass
600-mL beaker
500-mL graduated cylinder
clear container or 3 gallon bucket
water
chunk of wood
modeling clay
toy boat capable of a 1200-gram cargo, or 9-inch aluminum cake pan
100-g mass
3 lead masses or lead fishing weights

Discussion
An object submerged in water takes up space and pushes water out of the way. We say the water is *displaced*. Interestingly enough, the water that is pushed out of the way pushes back on the submerged object. For example, if the object pushes a volume of water with a weight of 100 N out if its way, then the water reacts by pushing back on the object with a force of 100 N — Newton's third law. We say that the object is *buoyed* upward with a force of 100 N. This is summed up in Archimedes' principle, which states that the *buoyant force* that acts on any completely or partially submerged object *is equal to the weight of the fluid the object displaces.*

Procedure
Step 1: Use a spring scale to determine the weight of an object (rock or hook mass) that is first out of water and then under water. The difference in weights is the buoyant force. Record.

Weight of object out of water = _____

Weight of object in water = _____

Buoyant force on object = _____

Step 2: Devise a method to find the volume of water displaced by the object. Record the volume of water displaced. Compute the mass and weight of this water. (Remember, 1 mL of water has a mass of 1 g and weighs 0.01 N.)

Volume of water displaced = _____

Mass of water displaced = _____

Weight of water displaced = _____

Boat Float

1. How does the buoyant force on the submerged object compare with the weight of the water displaced?

_____

_____

NOTE: To simplify calculations, for the remainder of this experiment measure and determine *masses,* without finding their equivalent *weights (W = mg).* Keep in mind, however, that an object floats because of a buoyant *force.* This force is due to the *weight* of the water displaced.

Step 3: Measure the mass of a piece of wood with a beam balance, and record the mass in Table A. Measure the volume of water displaced when the wood floats. Record the volume and mass of water displaced in Table A.

2. What is the relation between the buoyant force on any floating object and the weight of the object?

_____

_____

3. How does the mass of the wood compare to the mass of the water displaced?

_____

_____

4. How does the buoyant force on the wood compare to the weight of water displaced?

_____

_____

Step 4: Add a 100-g mass to the wood so that the wood displaces more water but still floats. Measure the volume of water displaced and calculate its mass, recording them in Table A.

5. How does the buoyant force on the wood with its 100-g load compare to the weight of water displaced?

_____

_____

Step 5: Roll the clay into a ball and find its mass. Measure the volume of water it displaces after it sinks to the bottom of a graduated cylinder. Calculate the mass of water displaced. Record all volumes and masses in Table A.

6. How does the mass of water displaced by the clay compare to the mass of the clay out of the water?

_____

_____

7. Is the buoyant force on the submerged clay greater than, equal to, or less than its weight out of the water? What is your evidence?

_____

_____

Step 6: Retrieve the clay from the bottom, and mold it into a shape that allows it to float. Sketch or describe this shape. Measure the volume of water displaced by the floating clay. Calculate the mass of the water, and record in Table A.

Table A

OBJECT	MASS (g)	VOLUME OF WATER DISPLACED (ml)	MASS OF WATER DISPLACED (g)
WOOD			
WOOD AND 50-g MASS			
CLAY BALL			
FLOATING CLAY			

Summing Up

8. Does the clay displace more, less, or the same amount of water when it floats as it did when it sank?

_____

_____

9. Is the buoyant force on the floating clay greater than, equal to, or less than the weight of the clay?

_____

_____

10. What can you conclude about the weight of an object and the weight of water displaced by the object when it floats?

_____

_____

11. Is the buoyant force on the clay ball greater when it is submerged near the bottom of the container or when it is submerged near the surface? What is your evidence?

_____

_____

12. Is the pressure that the water exerts on the clay ball greater near the bottom of the container than when submerged near the surface? Exaggerate the depth involved and cite expected evidence.

_____

Boat Float

13. Why are your last two answers different?

_____

Going Further

Step 7: Suppose you are on a ship in a canal lock. If you throw a ton of lead bricks overboard from the ship into the canal lock, will the water level in the canal lock go up, down, or stay the same? Write down your prediction *before* you proceed to Step 8.

Prediction for water level in canal lock: _____

Step 8: Float a toy boat loaded with lead "cargo" in a relatively deep container filled with water (deeper than the height of the lead masses). For observable results, the size of the container should be just slightly bigger than the boat. Mark and label the water levels on masking tape placed on the container and on the sides of the boat. Remove the masses from the boat and put them in the water. Mark and label the new water levels.

14. What happens to the water level *on the side of the boat* when you remove the cargo? What does this say about the amount of cargo carried by ships that float high in the water?

_____

15. What happens to the water level *in the container* when you place the cargo in the water? Explain why this happens.

_____

_____

16. Similarly, what happens to the water level in the canal lock when the bricks are thrown overboard?

_____

17. Suppose the freighter is carrying a cargo of Styrofoam instead of bricks. What happens to the water level in the canal lock if the Styrofoam (which floats in water) were thrown overboard?

_____

18. When a ship is launched at a shipyard, what happens to the sea level all over the world — no matter how imperceptibly?

_____

19. When a ship in the harbor launches a little rowboat from the dock, what happens to the sea level all over the world — no matter how imperceptibly?

_____

20. Cite evidence to support your (different?) answers to 18 and 19.

_____

CONCEPTUAL **Physical Science** ‎ **Experiment**

Force and Pressure

Tire Pressure and 18 Wheelers

Purpose
To distinguish between pressure and force.

Required Equipment and Supplies
automobile
owner's manual for vehicle (optional)
graph paper
tire pressure gauge

Discussion
People commonly confuse *force* and *pressure*. Tire manufacturers add
to this confusion by saying "Inflate to 45 pounds" when they really
mean "45 pounds *per* square inch." The pressure on your feet is
painfully more when you stand on your toes than when you stand on
your whole feet, even though the force of gravity (your weight) is
the same. Pressure depends on how a force is distributed. Pressure
increases as area decreases, and decreases as area increases.

$$\text{Pressure} = \frac{\text{force}}{\text{area}}$$

$$\text{Force} = \text{pressure x area}$$

Ever wonder why trucks that carry heavy loads have so many tires?
Count the tires on the trucks you see on the highway. Some have 18! Why?

Procedure
Step 1: Position a piece of graph paper directly in front of each tire of
an automobile. Roll the automobile onto the papers.

Step 2: Use a tire gauge to measure the pressure in each tire
in pounds per square inch. Record the pressures of each tire.
Trace the outline of the tire where it makes contact with the
graph papers. Roll the vehicle off the papers. Calculate
the area inside the trace in square inches or square cm.
Record your data in Table A.

Step 3: Study the tread pattern of the tires. Compare this to the tire marks on the papers.
Note that only the tread actually presses against the road. The gaps between the tread do not
support the vehicle. Estimate what fraction of the tire's contact area is tread and record it in
Table A. The area actually in contact with the road is the area inside your trace multiplied by
the fraction of tread. Repeat for all four tires.

Step 4: Compute the force each tire exerts against the road by multiplying the pressure
of the tire times its area of contact with the road. Record your computations in Table A.

Tire Pressure and 18 Wheelers

TIRE	OUTLINE TRACED ON PAPER (in²)	ESTIMATE OF % TREAD	AREA OF CONTACT (in²)	PRESSURE (lb/in²)	FORCE (lb)
Table A					
RIGHT FRONT					
LEFT FRONT					
RIGHT REAR					
LEFT REAR					

Step 5: Compute the weight of the vehicle by adding the forces from each tire.

   Computed weight = _____lb

Is this not quite remarkable — that the weight of a vehicle is equal to the air pressure in its tires multiplied by the area of tire contact? (See if your Uncle Harry knows this!)

Step 6: Find the weight of the vehicle from another source, such as the owner's manual or the local dealer. Sometimes this information is stamped on a plate on the inside of a door jamb.

   Known weight = _____lb

Summing Up

1. How does your computed value of vehicle weight compare to the stated value? What might account for a difference in the actual weight and the weight stated in the owner's manual?

_____

2. When you inflate tires to a higher pressure, what happens to the contact area of the tire against the road surface?

_____

3. A flat tire will read zero on a pressure gauge, when actually air at atmospheric pressure is inside. So a tire gauge is calibrated to measure the air pressure over and above atmospheric pressure. We distinguish between *gauge pressure*, and *atmospheric pressure*. Atmospheric pressure is normally 14.7 lb/in² or 10^5 N/m² outside the tire. Gauge pressure is the amount of pressure inside the tire over and above atmospheric pressure. So what is the *total* pressure inside each tire?

_____

4. Why was the atmospheric pressure (14.7 lb/in²) *not* added to the pressure of the tire gauge when computing the weight of the vehicle?

_____

_____

5. Why do trucks that carry heavy loads have so many wheels — often 18?

_____

_____

Tire Pressure and 18 Wheelers

CONCEPTUAL **Physical Science** **Experiment**

Temperature

Temperature Mix

Purpose
To predict the final temperature of a mixture of cups of water at different temperatures.

Required Equipment and Supplies
3 Styrofoam cups
liter container
thermometer (Celsius)
pail of cold water
pail of hot water
empty pail for waste water

Discussion
If we mix a pail of cold water with a pail of hot water, the final temperature of the mixture will be between the two initial temperatures. What
information would we need to predict the final temperature? We'll begin
with the simplest case of mixing *equal* masses of hot and cold water.

Procedure
Step 1: Begin by marking your three Styrofoam cups equally at about
the three-quarter mark. You can do this by pouring water from one cup
to the next and mark the levels along the inside of each cup.

Step 2: Fill the first cup to the mark with hot water from the hot pail, and fill the second cup
with cold water from the cold pail to the same level. Measure and record the temperature of
both cups of water.

Temperature of cold water = _____

Temperature of hot water = _____

Step 3: Predict the temperature of the water when the two cups
are mixed. Then pour the two cups of water into the liter
container, stir the mixture slightly, and record its temperature.

Predicted temperature = _____

Actual temperature = _____

1. If there was a difference between your prediction and your observation, what may have
   caused it?

_____

Be sure to pour the mixture into the sink or waste pail. (Don't be a clutz and pour it back into
either of the pails of cold or hot water!) Now we'll investigate what happens when *unequal*
amounts of hot and cold water are mixed.

Step 4: Fill one cup to its mark with cold water from the cold pail. Fill the other two cups to their marks with hot water from the hot pail. Measure and record temperatures of each cup. Predict the temperature of the water when the three cups are mixed. Then pour the three cups of water into the liter container, stir the mixture slightly, and record its temperature.

Predicted temperature = _____

Actual temperature of water = _____

Pour the mixture into the sink or waste pail. Again, do *not* pour it back into either of the pails of cold or hot water!

2. How did your observation compare with your prediction?

_____

3. Which of the water samples (cold or hot) changed more when it became part of the mixture? In terms of energy conservation, suggest a reason for why this happened.

_____

Step 5: Fill two cups to their marks with cold water from the cold pail. Fill the third cup to its marks with hot water from the hot pail. Measure and record their temperatures. Predict the temperature of the water when the three cups are mixed. Then pour the three cups of water into the liter container, stir the mixture slightly, and record the temperature.

Predicted temperature = _____

Actual temperature = _____

Pour the mixture into the sink or waste pail. (By now you and your lab partners won't alter the source temperatures by pouring waste water back into either of the pails of cold or hot water.)

4. How did your observation compare with your prediction?

_____

5. Which of the water samples (cold or hot) changed more when it became part of the mixture? Suggest a reason for why this happened.

_____

_____

Summing Up
6. What determines whether the equilibrium temperature of a mixture of two masses of water will be closer to the initially cold or hot water?

_____

7. How does the formula $Q = mc\Delta T$ apply here?

_____

_____

CONCEPTUAL **Physical Science** ————————————— **Experiment**

Specific Heat Capacities

Spiked Water

Purpose
To predict the final temperature of initially cold water after hot nails are added.

Required Equipment and Supplies
Harvard trip balance or equivalent
2 large insulated cups
bundle of short, stubby nails tied together with string
thermometer (Celsius)
hot and cold water
paper towels

Discussion
If you throw a hot rock into a pail of cool water, we know that the rock's temperature will decrease. We also know that the temperature of the water will increase — but will its increase in temperature be more, less, or the same as the temperature decrease of the rock? Will the temperature of the water go up as much as the temperature of the rock goes down? Or will the changes of temperature depend on the mass of rock and the mass of water?

Procedure
Step 1: Place a large cup on each pan of a beam balance. Place a bundle of nails into one of the cups. Add cold water to the other cup until it balances the cup of nails. When the two cups are balanced, the same mass is in each cup — a mass of nails in one, and an equal mass of water in the other. Visually determine if the amount of water in one cup is sufficient to cover the nails in the other cup. If not, add more nails to the bundle so that the corresponding mass of water will completely cover the nails.

Step 2: Set the cup of cold water on your work table. Remove the bundle of nails from its cup and place the cup beside the cup of cold water.

Step 3: Fill the empty cup 3/4 full with hot water. Lower the bundle of nails into the hot water. Be sure that the nails are completely covered by the hot water. Allow the nails and the water to reach thermal equilibrium.

Step 4: Measure and record the temperature of the cold water and the temperature of the hot water around the nails.

1. Is the temperature of the hot water equal to the temperature of the nails? Why do you think it is or is not? (Can you think of a better way to heat the nails to a known temperature?)

_____

_____

_____

Step 5: Before you dunk the hot nails into the cold water, predict the resulting temperature of the mixture. Then lift the nails from the hot water and put them quickly into the cold water. Be sure that the nails are completely covered by the cold water. When the temperature of the mixture stops rising, record it.

Predicted temperature = _____

Actual temperature = _____

2. How close is your prediction to the observed value?

_____

_____

Step 6: Now repeat Steps 1 through 5 with hot water replacing cold water and cold nails. First dry the bundle of nails with a paper towel. Then balance a cup with the dry bundle of nails with a cup of *hot* water. Remove the nails and fill the cup 3/4 full with *cold* water. Record the temperature of the hot water in the first cup. Lift the nails from the cold water and *quickly* lower the bundle of nails into the hot water. Be sure that the nails are completely covered. Predict the resulting temperature of the mixture before dunking the cold nails.

Predicted temperature = _____

Actual temperature = _____

Summing Up

3. How close is your prediction to the observed value?

_____

_____

4. Discuss your observations with your partners and write an explanation for what happened.

_____

_____

_____

5. Suppose you have cold feet when you go to bed, and you want something to warm your feet throughout the night. Would you prefer to have a bottle filled with hot water, or one filled with an equal mass of nails at the same temperature as the water? Explain.

_____

_____

_____

CONCEPTUAL **Physical Science** ——————————— **Experiment**

Specific Heat Capacities of Substances

Specific Heat Capacities

Purpose
To measure the specific heat capacities of some common metals.

Required Equipment and Supplies
hot plate
metal specimens
beaker
tongs
Styrofoam cups
balance
thermometer

Discussion
Have you ever held a hot piece of pizza by its crust only to have the moister parts burn your mouth when you take a bite? The meats and cheese have high specific heat capacity, whereas the crust has a low specific heat capacity. How can you compare the specific heat capacities of different materials?

In this experiment you will increase the temperature of metal specimens to that of boiling water. Then you'll place each specimen in a double Styrofoam-cup that contains a mass of room-temperature water equal to the mass of each specimen. The specimen will cool and the water temperature will rise until both are at the same temperature. That is, the heat lost by the specimen equals the heat gained by the water.

The specific heat c = the quantity of heat Q per mass m per change in temperature T

$$c = \frac{Q}{mc\,T}$$

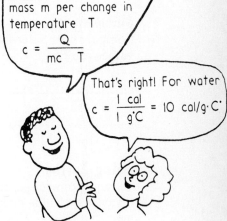

That's right! For water $c = \frac{1\ cal}{1\ g^{\circ}C} = 10\ cal/g\cdot C^{\circ}$

$$Q_{lost} = Q_{gained}$$
$$m_s c_s \Delta T_s = m_w c_w \Delta T_w$$

The specific heat capacity of the specimen is

$$c_s = \frac{m_w c_w \Delta T_w}{m_s \Delta T_s}$$

For water c_w = 1.00 cal/g °C and if the mass of water is the same as the mass of the specimen, then the specific heat of the sample is simply the ratio of the temperatures:

$$c_s = 1.00\,\tfrac{cal}{g\cdot^{\circ}C}\left(\frac{\Delta T_w}{\Delta T_s}\right)$$

Procedure
Step 1: Measure the mass of each specimen, and record it in Table A. Then place the specimens in a beaker of water, deep enough to cover the specimens, and heat the beaker to boiling.

Specific Heat Capacities

Step 2: While the beaker is being brought to a boil, assemble as many pairs of Styrofoam cups with one cup inside the other as you have specimens. You have just constructed inexpensive double-walled calorie meters, called *calorimeters*. Since 1 mL of water has a mass of 1 g, carefully measure as many mL of tap water as there are grams for each specimen and place the measured water in each calorimeter. Measure the temperature of this water in the calorimeters with a thermometer.

Step 3: Let the water in the beaker boil vigorously for more than a minute until you are convinced the specimens are in thermal equilibrium with the boiling water. Then using tongs, quickly remove each specimen from the boiling water, shake away excess droplets of water, and place each in the appropriate calorimeter (where the mass of contained water is the same as the mass of the specimen).

Step 4: After the nails have given their thermal energy to the water, record the final temperature of the water in each calorimeter — which is the same as the final temperature of each specimen. Enter your findings in Table A.

Table A

SPECIMEN	MASS (g)	$T_{Initial}$ (°C)	T_{final} (°C)	ΔT (°C)	C (cal/g °C)

Summing Up

Compare your values for the specific heats of your specimens to the table below. How do your values compare?

_____

_____

Going Further

Try an unknown specimen and see how closely it matches the value of one in the table of specific heat capacities.

Table B

Specific Heat Capacities	
Substance	c $\frac{cal}{g\text{-}C°}$
Aluminum	0.215
Copper	0.0923
Gold	0.0301
Lead	0.0305
Silver	0.0558
Tungsten	0.0321
Zinc	0.0925
Mercury	0.033
Water	1.000

Specific Heat Capacities

Name _____ Section _____ Date _____

Temperature of a Flame

Purpose
To measure the temperature of a flame using the concept of specific heat capacity.

Required Equipment and Supplies
calorimeter (2 Styrofoam cups with cardboard lid)
thermometer
metal tongs
20-gram brass weight
graduated cylinder (100 mL)
Bunsen burner
ring stand with wire gauze

Discussion
How hot is a flame? Hot enough to shatter any thermometer placed inside it–so don't! Instead, we can measure the temperature of a flame using an indirect approach. First, a piece of brass is heated by the flame and hence brought to about the same temperature. The brass is dropped into some water, which heats up the water. Knowing the specific heat capacity of water and of the brass, and measuring the temperature change of the water, we can calculate the initial temperature of the brass, and hence the temperature of the flame.

Safety
Remove all combustible materials from your work station when working with a Bunsen burner.
Be careful not to touch the brass weight while it is heated.

Procedure
Step 1: Add 100 mL of room-temperature water (100 g) to a Styrofoam cup tucked inside a second Styrofoam cup. Cover the container with the cardboard cover. Insert the thermometer through the hole in the cardboard and gently stir to assure a homogeneous temperature. Record this initial temperature of the water in Table 1.

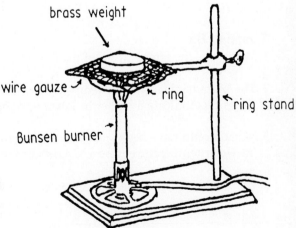

Step 2: Place a 20 g mass of brass on a wire gauze supported by a ring stand as shown in Figure 1. Ignite the Bunsen burner and adjust the height of the brass so that it is within the flame. Heat for a few minutes so that the brass reaches the same temperature as the flame. The brass may turn dark upon being heated.

Figure 1. Set-up for heating brass weight.

Step 3: Use tongs to pick up the brass and quickly drop it into the 100 g of water. Close the lid as soon as possible (and note the noise of bubbling!). Stir gently with the thermo-meter. Record the *maximum* temperature (Final Temperature) reached in the table shown.

Temperature of a Flame

Mass of...	Initial Temperature (°C)	Final Temperature (°C)
water:		
brass:		

Step 4: The final temperature of the water and the brass are the same. Pick the brass out of the water with your bare bands to prove this to yourself. Note that like the water it is also lukewarm. The only unknown in Table 1 at this point is the initial temperature of the brass, which is the very high temperature *immediately* before the brass was added to the water. Perform the following calculations to find this initial temperature of the brass.

Calculations

The quantity of heat Q that changes the temperature ΔT of a mass m and specific heat capacity c is given by $Q = mc\Delta T$. For water, $c = 1.00$ cal/g°C. Enter the appropriate data into this equation to find the quantity of heat *gained* by the water: The units of your answer should be given in calories.

$Q_{water} =$ (_____ g)(1.00 cal/g°C)(_____°C – _____°C) = _____
  mass of water T_{final} $T_{initial}$ Heat gained by
                    of the water of the water   water

Heat was gained by the water, but where did this heat originate? It came from the brass. Ideally, the quantity of heat gained by the water is equal to the quantity of heat lost by the brass ($Q_{water} = - Q_{brass}$). We use the same equation for heat lost by the brass, where for brass, $c = 0.0917$ cal/g°C.

$Q_{brass} =$ (_____ g)(0.0917 cal/g°C)(_____°C – _____°C) = _____
  mass of brass $T_{initial}$ T_{final} Heat lost by
                    of the brass of the brass   brass

Since $Q_{water} = - Q_{brass}$ the only unknown in this second equation is the initial temperature of the brass. Solve for this with a bit of algebra. If needed, consult your lab partner for help.

Summing Up

1. What was the temperature of the flame? _____

2. The brass tends to change temperature much faster than the water.
   Does this indicate a higher or lower specific heat capacity? _____

3. What would have been your calculated flame temperature if the change
   in water temperature were 1°C greater than what you actually measured? _____

4. What are some possible sources of error for this experiment? _____

_____

Temperature of a Flame

CONCEPTUAL **Physical Science**   **Experiment**

Change of Phase

Taking the Heat

Purpose
To study the effect of adding heat to various phases of water.

Required Equipment and Supplies
aluminum can with top removed
aluminum foil
thermometer
3" x 3" cardboard piece with thermometer hole through the center
clamp
wire gauze, ring, and ring stand
Bunsen burner
graduated cylinder (250 mL)
balance

Discussion
Adding heat to water does not necessarily raise its temperature. There are other things that can happen as is investigated in this experiment.

Safety
Remove all combustible materials from your work station when working with a Bunsen burner. Do not let the aluminum foil used in this procedure get too close to the flame for it too can burn. Exercise caution when working with boiling water.

Procedure
Step 1: Add about half a liter of water to enough crushed ice to make a slurry. After stirring the slurry, measure 250 mL of strained ice-cold water into a 12 oz aluminum can, which is wrapped in aluminum foil and secured by a clamp over a wire gauze and Bunsen burner (See Figure 1). For convenience, it's best to use a can that has its top removed. Allow a few ice chips to pass into the can to assure that the temperature stays as close to 0°C as possible. Cover the can with the cardboard and insert the thermometer. Carefully stir the water with the thermometer to assure that the temperature of the water is homogeneous. Record the initial temperature in Table 1. Note that this temperature may not be exactly 0°C.

Step 2: Light the Bunsen burner and measure the temperature of the water every 30 seconds until it begins to boil. Let the water boil for a few readings then stop. Do not let the thermometer touch the bottom of the can while the water is being heated as the aluminum will conduct heat directly to the thermometer. Remember to keep stirring. Record all temperature readings in Table 1.

Step 3: Repeat steps 1 and 2 using 250 grams of crushed ice rather than 250 mL of ice cold water.

Step 4: Plot your data using a color coded bar graph.

Taking the Heat

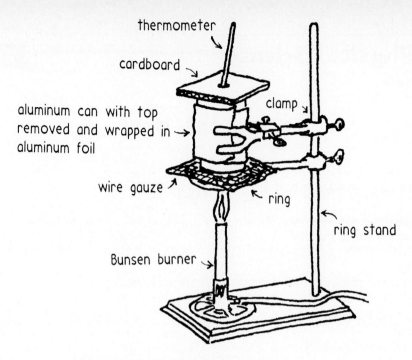

Figure 1 Set-up for applying heat to water.

Table 1

| | Temperature °C | |
TIME	Ice-Cold Water	Crushed Ice
0 minutes 0 seconds		

(continued)

Taking the Heat

| | Temperature | |
Time	Ice-Cold Water	Crushed Ice

Summing Up

1. The heat added to ice-cold water went directly to increasing the motion (kinetic energy) of the water molecules. This you observed as an increase in temperature. What happened to the initial heat added to the crushed ice?

2. What do you suppose had been occurring just before the temperature started to markedly increase inside the can containing crushed ice?

3. Why did the temperature not increase as heat was added to the boiling water?

4. Is it possible to add heat to ice without the ice melting? Briefly explain.

Taking the Heat

CONCEPTUAL **Physical Science**   **Experiment**

● *Change of Phase*

Cooling by Boiling

Purpose
To experience that water may boil only because air pressure is being lowered.

Required Equipment and Supplies
400 mL beaker
thermometer
vacuum pump with bell jar

Discussion
Evaporation is a change of phase from liquid to gas at the surface of a
liquid; boiling is a rapid change of phase from liquid to gas at and below
the surface of a liquid. The temperature at which water boils depends on
atmospheric pressure. Have you ever noticed that water reaches its
boiling point in a *shorter* time when camping up in the mountains? And
have you noticed that at high altitude it takes *longer* to cook potatoes
or other food in boiling water? That's because when the pressure of the
atmosphere on its surface is reduced, water boils at a lower temperature.
Let's see!

Can it really be boiling?

Procedure
Warm 200 mL of water in a 400 mL beaker to a temperature above 60°C. Record the temperature.
Then place the beaker under and within the bell jar of a vacuum pump. If a thermometer will fit
underneath the bell jar, place a thermometer in the beaker. Turn on the pump. What happens to
the water?

T = _____

_____

_____

Was the water *really* boiling?

_____

_____

Stop the pump and remove the bell jar.
What is the temperature of the water now?

T = _____

As time permits, repeat the procedure, starting with other temperatures, such as 80°C, 40°C, and
20°C, recording the time it takes for boiling to begin.

Summing Up

1. In terms of energy transfer, what does it mean to say that boiling is a cooling process? What cools? What warms?

_____

_____

_____

2. Name two ways to cause water to boil.

_____

_____

3. Boiling water on a hot stove remains at a constant 100°C temperature. How is this observation evidence that boiling is a cooling process?

_____

_____

_____

_____

CONCEPTUAL **Physical Science** ———————— **Experiment**

Change of Phase

Warming by Freezing

Purpose
To experience that heat is released when crystallization (freezing) occurs.

Required Equipment and Supplies
RE-HEATER® packs
hot plate
large pan or pot of boiling water
large Styrofoam cups (600 mL, if available)

Discussion
We know that it is necessary to add heat to liquefy (melt) a solid or gasify (evaporate, vaporize, or boil) a liquid. In the reverse process, heat is released when a gas liquifies (condenses) or a liquid solidifies (freezes). Thermal energy that accompanies these changes of phase is called *latent heat of vaporization* (going from gas to liquid or liquid to gas), and *latent heat of fusion* (going from liquid to solid or solid to liquid).

Energy is absorbed when change of phase is in this direction

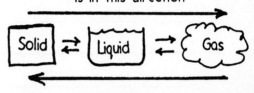

Energy is released when change of phase is in this direction

Water, which at standard pressure normally freezes at 0°C (32°F), can be found under certain conditions in a liquid phase as low as –40°C (–40°F). This *supercooled* water (liquid water below 0°C) often exists as tiny cloud droplets, common in clouds where snow or ice particles form. Freezing in clouds depends on the presence of *ice-forming nuclei*, most of which are active in the –10°C to –20°C range. Ice-forming nuclei may be many different substances such as dust, bacteria, other ice particles, or silver iodide used to "seed" clouds during droughts. Silver iodide is active at temperatures as high as –4°C.

Formation of ice particles can be dangerous to aircraft. The skin of the aircraft, well below the freezing temperature, provides an excellent surface on which supercooled water suddenly freezes. This is called aircraft icing, which can be quite severe under certain conditions.

The RE-HEATER provides a dramatic simulation of a supercooled liquid. What you observe in the RE-HEATER is actually the release of the latent heat of crystallization, which is analogous to the release of latent heat of vaporization or the latent heat of fusion.

It only takes a quick click of an internal metal disk to activate the RE-HEATER. The internal disk has two distinct sides. If you use your thumb and forefinger and squeeze quickly, you will not need to worry about which side is up.

After observing the crystallization of the sodium acetate and the heat released, you might want to try it again and measure the heat of crystallization. The package has a mass of about 146 grams. The packaging and the trigger mechanism have a mass of about 26 grams. Thus, the sodium acetate solution inside the package has a mass of about 120 grams.

Warming by Freezing

Procedure

Place the RE-HEATER in an insulated container (such as a larger Styrofoam cup) with 400 g of room-temperature water. Allow the water and RE-HEATER to reach equilibrium. Measure and record its initial temperature. Activate the RE-HEATER by clicking the metal disk. Immediately return the RE-HEATER to the container and measure the temperature of the water at regular intervals. How much heat is gained by the water?

$Q =$ _____

Summing Up

1. How does the crystallization inside the RE-HEATER relate to heating and air conditioning in a building? (*Hints:* Think about how steam heat and radiators operate or how the refrigerants in an air conditioner operate.)

_____

_____

_____

2. Speculate about how these processes relate to airplane safety.

_____

_____

_____

3. Think of and list some practical applications of the RE-HEATER.

_____

_____

_____

Wait a minute — before this we cooled by boiling, and now we warm by freezing? Must be some interesting physics going on here!

CONCEPTUAL **Physical Science** ⌐ Activity ⌐

Electrostatics

Charging Up

Purpose
To observe the effects and behavior of static electricity.

Required Equipment and Supplies
two new balloons
Van de Graaff generator
matches
Styrofoam peanuts or fresh puffed rice

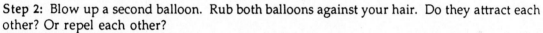

Discussion
Skuff your feet across a rug and reach for a doorknob and zap — electric shock! The electric potential that causes the spark can be several thousand volts, which is why technicians have to be so careful when working with tiny circuits such as those in computer chips!

Procedure
Step 1: Blow up a balloon. After stroking it against your hair, place it near some small pieces of Styrofoam or puffed rice. Then place the balloon against the wall where it will "stick," as shown above right. On the drawing, sketch the arrangement of some sample charges on the balloon and on the wall.

Step 2: Blow up a second balloon. Rub both balloons against your hair. Do they attract each other? Or repel each other?

Step 3: Stand on an isolation stand (or rubber mat) next to a discharged Van de Graaff generator. Place one hand on the conducting sphere on top of the generator and have your partner switch on the generator motor. Shake your head as the generator charges. What do you experience?

Step 4: Discharge the generator by touching it with your finger or knuckle. Place a small cup of puffed rice or Styrofoam chips on top of the conducting sphere. What happens when you turn on the generator?

Step 5: Light a wooden match and move it near a charged sphere on top of the generator. What do you observe?

Opposites attract ...
Likes repel...

Summing Up
Offer an explanation for your observations of Steps 1 to 5 in terms of the fundamental rule of electricity — *like charges repel and unlike charges attract.*

_____

_____

Charging Up

CONCEPTUAL **Physical Science** ———————————— | **Activity** |
Electric Circuits

Batteries and Bulbs

Purpose
To explore various arrangements of batteries and bulbs, and the effects of those arrangements on bulb brightness (intensity).

Required Equipment and Supplies
3 size-D dry-cells
about 50 cm of bare copper wire
3 flashlight bulbs (1.5 volt)
3 porcelain or plastic bulb sockets

Discussion
Many devices include electronic circuitry, most of which are quite complicated. Complex circuits are made, however, from simple circuits. In this activity we build one of the simplest yet most useful circuits ever invented — that for lighting a light bulb! To the right are common elements found in electric circuits. It is common to call a single cell a battery, but strictly speaking, a battery refers to a combination of cells. So instead of saying "a battery of cells," we simply say "a battery."

Procedure
Step 1: Arrange one bare bulb, one cell, and connecting wire in as many ways as you can to make the bulb light up. On a separate sheet of paper, sketch each of your arrangements — failures as well as successes.

Step 2: Using a bulb in a bulb socket (instead of a bare bulb), one cell, and a piece of wire, try lighting the bulb in as many ways as you can. Sketch your arrangements and note the ones that work.

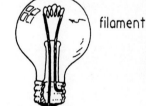
filament

Step 3: Using one battery, light as many bulbs in sockets as you can. Sketch your arrangements and note the ones that work. If possible, compare your results with those of other students.

Step 4: Connect the bulbs, battery, and wire as shown in Figure A. Circuits like these are examples of *series* circuits.

Step 4: Set up the circuit shown in Figure B. A circuit like this is a *parallel* circuit.

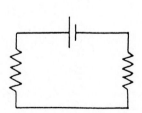

Fig. A Simple series circuit

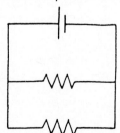

Fig. B Simple parallel circuit

Batteries and Bulbs

Summing Up

1. With what two parts of the bulb does the bulb holder make contact?

_____

_____

2. What do successful arrangements of batteries and bulbs have in common?

_____

_____

3. How does the brightness of a pair of bulbs compare when connected in series in these two circuits?

_____

_____

4. How does the brightness of the bulbs in this circuit compare to that in the series circuit?

_____

_____

5. What happens if you unscrew a bulb in either circuit?

_____

_____

6. How do you suppose most of the circuits in your home are wired — in series or in parallel? What is your evidence?

_____

_____

CONCEPTUAL **Physical Science**

Electric Circuits

3-Way Switch

Purpose
To explore ways to turn a light bulb on or off from either of two switches.

Required Equipment and Supplies
2.5 volts DC light bulb with sockets
hook-up wire
two 1.5-volt size D cells with holder, connected in series
two single pole double-throw switches

Discussion
A multi-storey home, or an industrial building, often has stairways with a ceiling light. It is convenient to turn the ceiling light on or off using a switch at the bottom of the stairway, and also to be able to turn the light on or off using a switch located at the top of the stairway. Each switch should be able to turn the light on or off, regardless of the previous setting of either switch. In this activity you will see how simple but tricky such a common circuit really is!

Procedure
Step 1: Connect a wire from the positive terminal of a 3 volt battery (two 1.5 volt cells connected positive-to-negative) to the center terminal of a single-pole double-throw switch.

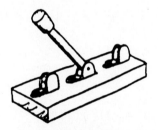

Single pole
double-throw switch

Step 2: Connect a wire from the negative terminal of the same battery to one of the light bulb terminals. Connect the other light bulb terminal to the center terminal of the second switch.

Step 3: Now interconnect the free terminals of the switches so that the bulb turns on or off from either switch — regardless of the setting of the other switch.

Step 4: When you succeed, draw a simple circuit diagram of your successful circuit.

Summing Up
Will your successful circuit work if you reverse the polarity of the battery? What is your evidence?

_____

_____

CONCEPTUAL **Physical Science** ┌─────────┐
                                │ Activity │
                                └─────────┘

Magnetism

Magnetic Personality

Required Equipment and Supplies
2 bar magnets
strong horseshoe magnet
jar of iron filings in oil
dry iron filings
paper
several feet of copper wire
galvanometer or milliammeter
small compass

Purpose
To explore the shape of a magnetic field, and the influence of a moving magnet in a coil of wire — electromagnetic induction.

Discussion
A magnetic field is a kind of aura that surrounds magnets. Although it can't be seen directly, the overall shape of the field can be seen by the effect it has on iron filings.

Procedure
Step 1: Place a bar magnet on a horizontal surface and cover it with a sheet of paper. Sprinkle iron filings on the paper. Jiggle it a bit and see the magnetic field shape. Repeat for a horseshoe shaped magnet. Sketch the indicated field for the bar magnet and horseshoe magnet on the drawings below.

Step 2: Place a pair of bar magnets on the table and repeat Step 1. The field of one magnet interacts with the field of the other to produce some interesting effects. Be sure to arrange the magnets in at least two configurations: like poles facing each other; and unlike poles facing each other. Sketch the indicated field for the two magnet configurations on the drawings below.

Step 3: Vigorously shake the jar of iron filings. Place a strong horseshoe magnet against the jar and observe carefully. Try different locations and see how the filings align. How do the filing orientations differ from those in the previous steps?

_____

_____

Step 4: Attach the ends of a long length of wire to the terminals of a sensitive galvanometer or sensitive milliammeter. Make several loops in the wire, as suggested in the figure, and plunge a strong bar magnet in and out of the loops. Try magnets of different strengths. Investigate the difference in plunging the north pole of the magnet in the loop, versus plunging the south pole in.

Optional
Place a magnet in front of a TV picture. (Do this with a TV you don't value. If you do it with a color set, you may permanently mess it up!)

Summing Up

1. How did the magnetic field lines in Step 2 compare for attracting magnets and repelling magnets?

_____

_____

2. Under what conditions did you generate the largest current in Step 4?

_____

_____

3. How is Step 4 related to the metal detectors used for detecting iron items on passengers in airline terminals?

_____

_____

Magnetic Personality

CONCEPTUAL **Physical Science**

Electric Circuits

Cranking Up I

Purpose
To compare qualitatively the power inputs to a series and to a parallel circuit.

Required Equipment and Supplies
hand crank generator (Genecon) or equivalent
parallel bulb apparatus

Discussion
Here's a chance to both see and feel some differences between series and parallel circuits.

Procedure

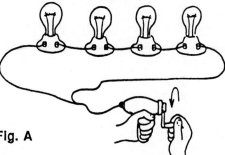

Step 1: Assemble four bulbs in a series configuration as shown in Figure A. Screw all the bulbs into their sockets. Connect the sockets with clip leads or wires.

Step 2: Connect the two leads of a Genecon hand-cranked generator to the ends of the string of bulbs. Crank the Genecon so that all the bulbs light up. **Fig. A** Now disconnect one of the bulbs from the string and re-connect the Genecon. Crank the Genecon so the three remaining bulbs are energized to the same brightness as the four-bulb arrangement. How does the crank *feel* now? Repeat, removing one bulb at a time, comparing the cranking action each time.

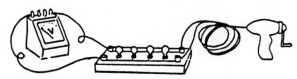

Step 3: Assemble the circuit with the parallel-bulb apparatus as shown in Figure B. Each end of the bulb apparatus has two terminals. Connect the lead of a voltmeter to one pair of terminals on one end of the apparatus. **Fig. B**

Step 4: Connect the leads of the Genecon to the terminals on the other end of the apparatus. Crank the Genecon with all the bulbs unscrewed in the sockets so they don't light. Then have your partner screw them in one at a time as you crank on the Genecon. Try to keep the bulbs energized at the same brightness as each bulb is screwed into its socket.

Summing Up
1. What do you notice about the *effort* required to crank the Genecon at a constant speed as more bulbs were added to the parallel circuit?

_____

2. How would you compare the amount of effort required to crank the Genecon to energize four bulbs in series with the effort to energize to four bulbs in parallel?

_____

CONCEPTUAL Physical Science———————————————— `Experiment`
Electric Circuits

Cranking Up II

Purpose
To compare quantitatively the power inputs to a series circuit and to a parallel circuit.

Required Equipment and Supplies
hand-crank generator (Genecon preferred)
four light bulbs with sockets
three 1.5-volt cells
parallel bulb apparatus
voltmeter
ammeter

Discussion
Here we repeat *Cranking-Up I*, which was a qualitative activity, but this time use a voltmeter and ammeter to make a quantitative comparison. For numerical relationships, we'll use batteries in place of the hand-crank generator.

Procedure
Step 1: Assemble four bulbs in a series circuit and connect the meters as shown in Figure A. Connect the voltmeter in parallel with the bulbs so you can measure the total voltage applied to the circuit as well as the voltage across each bulb. Play around with the Genecon, and then to provide a more constant voltage, use a 3-volt source, by connecting a pair of 1.5-volt cells in series. Connect one end to one terminal of the bulbs and the ground connection to one lead of an ammeter. Connect the other lead of the ammeter to the other terminal of the bulbs. The ammeter will measure the *total* current in the circuit.

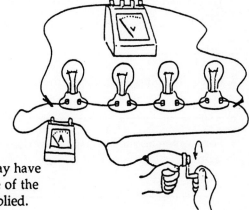

Fig. A For quantitative measurements, replace the Genecon, shown here, with a battery of cells — first 3 volts, then 4.5 volts.

Note: If you are *not* using digital meters, you may have to reverse the polarity of the leads if the needle of the meter goes the wrong way (–) when power is applied.

Close the switch and apply power to the circuit and then measure:

      a) the current in the circuit.
      b) the voltage applied to the circuit.
      c) the voltage across each bulb.

Now shortcircuit the bulbs in turn, so you'll investigate and compare circuits of 4 bulbs, then 3, then 2, and finally 1 bulb. Record your data in Table A.

Table A

# OF BULBS	TOTAL CURRENT (A)	TOTAL VOLTAGE (V)	VOLTAGE ACROSS EACH BULB (V)			
1						
2						
3						
4						

Step 2: Add another battery to your 3-volt supply so you have a 4.5-volt supply. Repeat for this higher voltage and record your data in Table B.

Table B

# OF BULBS	TOTAL CURRENT (A)	TOTAL VOLTAGE (V)	VOLTAGE ACROSS EACH BULB (V)			
1						
2						
3						
4						

Summing Up (Series Circuit)

1. What changes in brightness, if any, did you observe with changing numbers of bulbs in the circuit?

_____

2. Does the voltage applied in the circuit change as you add more bulbs? What is your evidence?

_____

3. For a given applied voltage, how does the current in the circuit change when more bulbs are added?

_____

4. Did any of the relationships you discovered between voltages and currents change when you applied 4.5 volts instead of 3 volts? What is your evidence?

_____

_____

Step 3: Assemble the circuit and connect the meters as shown in Figure B. Connect the voltmeter in parallel with the bulbs by connecting the voltmeter to two terminals on one end of the parallel-bulb apparatus. Connect the 3-volt lead from the voltage supply to one terminal of the parallel-bulb apparatus. Connect the ground lead from the voltage supply to one lead of an ammeter; connect the other lead of the ammeter to the second terminal of the parallel-bulb apparatus. The ammeter will measure the *total* current in the circuit.

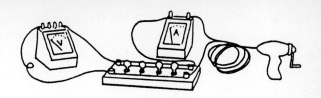

Fig. B For quantitative measurements, replace the genecon, shown here, with a battery of cells — first 3 volts, then 4.5 volts.

Make sure the bulbs are not loose in their sockets. Close the switch and apply power to the circuit. Observe their brightness. Now unscrew the bulbs one at a time.

Step 4: Screw the bulbs back in, one at a time, each time measuring:

    a) the current in the circuit.
    b) the voltage applied to the circuit.
    c) the voltage across each bulb.

Record your data in Table C.

Table C

# OF BULBS	TOTAL CURRENT (A)	TOTAL VOLTAGE (V)	VOLTAGE ACROSS EACH BULB (V)			
1						
2						
3						
4						

Step 5: Repeat Steps 3 and 4 using a 4.5-voltage supply. Record your data in Table D.

Table D

# OF BULBS	TOTAL CURRENT (A)	TOTAL VOLTAGE (V)	VOLTAGE ACROSS EACH BULB (V)			
1						
2						
3						
4						

Cranking Up II

Summing Up (Parallel Circuit)

6. Did you observe any change in bulb brightness as the number of bulbs in the circuit was changed? Briefly describe your observations.

_____

_____

7. Does the voltage across each bulb change as more bulbs are added to or subtracted from the circuit? What is your evidence?

_____

_____

8. How does the current in the battery pack change as the number of bulbs in the circuit changes?

_____

_____

9. Did any of the relationships you discovered between voltage and current change when you applied 4.5 volts instead of 3 volts? What is your evidence?

_____

_____

10. Consider lighting a pair of side-by-side identical bulbs. For maximum brightness of the 2-bulb set, should you connect them in series or connect them in parallel? Does it make any difference? What is your evidence?

_____

_____

CONCEPTUAL **Physical Science**

Vibrations and Sound

Slow-Motion Tuning Forks

Purpose
To observe and explore oscillations of a tuning fork.

Required Equipment and Supplies
variety of tuning forks (low frequency [40-150 Hz] forks work well for large amplitudes) strobe light, variable frequency

---

Discussion
The tines of a tuning fork oscillate at a very precise frequency. That's why musicians use them to tune instruments. In this activity we investigate their motion with a special illumination system — a stroboscope.

Procedure
Strike a tuning fork with a mallet or the heel of your shoe (do NOT strike against the table or other hard object). Does it appear to vibrate? Repeat, and this time immerse the tip of the tines just below the surface of water in a beaker. What do you observe?

_____

Now dim the room lights and strike a tuning fork while it is illuminated with a strobe light. For best effect, use the tuning fork with the longest tines available. Adjust the frequency of the strobe so that the tines of the tuning fork appear to be stationary. Then carefully adjust the strobe so that the tines slowly wag back and forth. Describe your observations.

_____

_____

Summing Up
1. What happens to the air next to the tines surface as they oscillate?

_____

2. Strike a tuning fork and observe how long it vibrates. Repeat placing the handle against the table top or counter. Although the sound is louder, does the *time* the fork vibrates increase or decrease? Explain?

_____

_____

3. Imagine you struck the tuning fork in outer space — what would happen then?

_____

Slow-Motion Tuning Forks

CONCEPTUAL **Physical Science**

Speed of Sound

Little Sir Echo

Purpose
To estimate the speed of sound using an echo.

Required Equipment and Supplies
large building with a flat wall
stopwatch or wrist watch with second hand

Discussion
This activity is similar to one Robert Millikan (the first American physicist to win the Nobel prize) performed when he measured the echo time between buildings with his class. He also did a similar experiment to measure the speed of light. We'll do this much simpler version.

Procedure
Find a flat side of a building (or a long corridor) with a good echo, and have your partner clap hands until you can hear successive echoes clearly. Position yourself so that you can measure the time from the clap to the last echo you can distinctly hear.

One method of doing this is to clap steadily, adjusting your rate until each reflection is heard exactly midway between the preceding and following claps. Once you get the rhythm, with a stopwatch measure the time it takes to clap 10 times.

The distance traveled during that time will be the number of echoes multiplied by the round-trip distance to the wall or end of the corridor. Estimate the speed of sound by dividing this distance by the time.

$v =$ _____

Summing Up
How close are you to the accepted value of 340 m/s at 20°C?

_____

_____

CONCEPTUAL **Physical Science** **Activity**

Wave Interference

Sound Off

Purpose
To dramatically demonstrate the interference of sound.

Required Equipment and Supplies
Stereo radio, tape, or CD player with two moveable speakers, one with a DPDT (double pole double throw) switch or a means of reversing polarity.

Discussion
Interference is a main property of all waves. With water waves we see it in regions of calm where overlapping crests and troughs coincide. We see the effects of interference with light in the colors of soap bubbles and other thin films where reflection from closeby surfaces puts crests coinciding with troughs. In this activity we'll dramatically experience the effects of interference with sound!

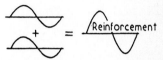

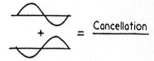

Procedure
Play the stereo player with both speakers in phase (with the plus and minus connections to each speaker the same). Play it in mono mode so the signals of each speaker are identical. Note the fullness of the sound. Now reverse the polarity of one of the speakers (either by physically interchanging the wires or by means of the switch provided). Note the sound is different — it lacks fullness. Some of the waves from one speaker are arriving at your ear out of phase with waves from the other speaker.

Now place the speakers facing each other at arm's reach distance apart. The long waves are interfering destructively, detracting from the fullness of the sound. Gradually bring the speakers closer to each other. What happens to the volume and fullness of the sound heard? Bring them face to face against each. What happens to the volume now?

Summing Up
1. What happens when to the volume of sound when the face-to-face speakers are switched so both are in phase?

_____

2. Why is the volume so diminished when the out-of-phase speakers are brought together face to face? And why is the remaining sound so "tinny"?

_____

3. What practical applications can you think of for canceling sound?

_____

CONCEPTUAL **Physical Science**

Properties of Light — Image Formation

Pinhole Image

Purpose
To investigate the operation of a pinhole "lens" and compare it to the eye.

Required Equipment and Supplies
3" x 5" card
straight pin
meter stick

Discussion
The image cast through a pinhole in a pinhole camera has the property of being in clear focus at any distance from the pinhole. That's because the tinyness of the pinhole does not allow overlapping of light rays. (The tinyness also doesn't allow the passage of much light, so pinhole images are normally dim as well.) When a pinhole is placed at the center of the pupil of your eye, the light that passes through the pinhole forms a focused image no matter where the object is located. Pinhole vision, although dim, is remarkably clear. In this activity, you will use a pinhole to see fine detail more clearly.

Procedure

Step 1: Bring this printed page closer and closer to your eye until you cannot clearly focus on it any longer. Even though your pupil is small, your eye does not act like a true pinhole camera because it does not focus well on nearby objects.

Step 2: Poke a single pinhole into a piece of card. Hold the card in front of your eye and read these instructions through the pinhole. Bright light on the print may be required. Bring the page closer and closer to your eye until it is a few centimeters away. You should be able to read the type clearly. Then quickly remove the card and see if you can still read the instructions without the benefit of the pinhole.

Summing Up
Enlist the help of people in your lab who are nearsighted and who are farsighted (if you're not one of them yourself).

1. A farsighted person without corrective lenses cannot see close-up objects clearly. Can a farsighted person without corrective lenses see close-up objects clearly through a pinhole?

_____

2. A nearsighted person without corrective lenses cannot see far-away object clearly. Can a nearsighted person without corrective lenses see far-away objects clearly through a pinhole?

_____

3. Why is a page of print dimmer when seen through the pinhole?

_____

Pinhole Image

CONCEPTUAL **Physical Science** ────────────────── **Experiment**

Properties of Light — Image Formation

Pinhole Camera

Purpose
To observe images formed by a simple convex lens and compare cameras with and without a lens.

Required Equipment and Supplies
covered shoebox
25 mm converging lens
glassine paper
aluminum foil
masking tape

Discussion
The first camera used a pinhole opening to let light in. Due to the smallness of the hole, light rays that entered could not overlap. This is why a clear image formed on the inner back wall of the camera. Because the opening was small, a long time was required to expose the film sufficiently. A lens allows more light to pass through and still focus the light onto the film. Cameras with lenses require much less time for exposure than pinhole cameras do, so their pictures came to be called "snapshots".

Procedure
Step 1: Construct a camera as shown in Figure A. It is a shoebox with a hole about an inch or so in diameter on one end, some glassine paper taped in the center to act as a screen, and an opening for viewing the screen on the other end. Tape some foil over the lens hole of the box. Poke a pinhole in the middle of the foil. Point the camera toward a brightly illuminated scene, such as the window during the daytime. Light enters the pinhole and falls on the glassine paper. Observe the image of the scene on the glassine paper.

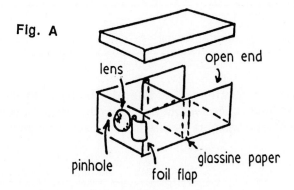

Fig. A

1. Is the image on the screen upside down (inverted)?

_____

2. Is the image on the screen reversed left to right?

_____

Step 2: Now remove the pinhole foil and tape a lens over the hole in the box. You now have a lens camera. Move it around and watch people or whatever.

3. Is the image on the screen upside down (inverted)?

_____

4. Is the image on the screen reversed left to right?

_____

Step 3: Unlike a lens camera, pinhole cameras focuses equally well on objects at practically all distances. Aim the camera lens at a nearby objects and see if the lens focuses on nearby objects.

5. Does the lens focus on nearby objects as well as it does on distant ones?

_____

Step 4: Draw a ray diagram as follows. First, draw a ray for light that passes from the top of a distant object through a pinhole and onto a screen. Second, draw another ray for light that passes from the bottom of the object through the pinhole and onto the screen. Then sketch the image created on the screen by the pinhole.

Summing Up
6. Why is the image created by the pinhole dimmer than the one created by the lens?

_____

_____

7. How is a pinhole camera similar to your eye? Do you think that the images formed on the retina of your eye are upside down? Your explanation might include a diagram.

_____

_____

8. With lens removed, punch 3 or 4 additional holes near the first pinhole. How many images are produced? Predict what would happen to the images if you placed the lens over the pinholes. Then do so and note the effect! Did you confirm your prediction, or were you surprised?

_____

_____

CONCEPTUAL **Physical Science**

Reflection

Mirror, Mirror, on the Wall . . .

Purpose
To investigate the minimum size mirror required for you to see a full image of yourself.

Required Equipment and Supplies
large mirror, preferably full length
ruler and masking tape

Discussion
Why do shoe stores and clothier shops have full-length mirrors? Need a mirror be as tall and wide as you for you to see a complete image of yourself?

Procedure
Step 1: Stand about arm's length in front of a vertical full-length mirror. Reach out and place a small piece of masking tape on the image of the top of your head. Now stare at your toes. Place the other piece of tape on the mirror where your toes are seen. Use a meter stick to measure the distance from top of your head to your toes. How does the distance between the pieces of tape on the mirror compare to your height?

_____

Step 2: Now stand about 3 meters from the mirror and repeat. Stare at the top of your head and toes and have an assistant move the tape so that the pieces of tape mark where head and feet are seen. Move further away or closer, and repeat. What do you discover?

_____

Summing Up
1. What is the shortest mirror in which you can see your entire image? Do you *believe* it?

_____

2. Does the location of the tape depend on your distance from the mirror?

_____

Going Further
Try this one if a full-length mirror is not readily available *or* you are a disbeliever! Hold a ruler next to your eye. Measure the height of a common pocket mirror. Hold the mirror in front of you so that the image includes the ruler. How many centimeters of the ruler appear in the image? How does this compare to the height of the mirror? Does distance affect your answer?

_____

CONCEPTUAL **Physical Science** | **Activity**

Polarized Light

Polaroids Ho!

Purpose
To investigate the effects of polarized light.

Required Equipment and Supplies
3 small polarizing filters
light source

Discussion
Light travels in waves, the vibrations of which are normally in random directions — in many planes at once (because the vibrating electrons that emit the light vibrate in random directions). In polarized light, the light waves vibrate in one plane only. Polarized light can be made by blocking all the waves except those in one plane. This is commonly done with polarization filters. The filters can also be used to detect polarized light.

Procedure
Step 1: Place one polarization filter between your eyes and a light source. Slowly rotate the filter a complete turn. Observe the intensity of the light as seen through the filter. Does the intensity change as you rotate the filter?

_____

Step 2: Place one filter in a fixed position in front of the light source. Slowly rotate a second filter held between your eyes and the fixed filter. Does the intensity change as you rotate the filter?

_____

Step 3: Hold the filter to your eye in a fixed position and have your lab partner rotate the other filter next to the light source. Any difference?

_____

Step 4: View different regions of the sky on a sunny day through a filter. Does the intensity change as you rotate the filter? Are parts of the sky polarized?

_____

_____

Step 5: View a liquid crystal display (LCD), such as on a wristwatch or calculator, through a polarization filter. Does the display change as you rotate the filter? Are liquid crystals polarized?

_____

Polaroids Ho!

Going Further

Step 6: Place a pair of filters in crossed position, so no light gets through. Place a third filter between the light source and the crossed filters. Does light get through?

_____

Step 7: Then place the third filter between the crossed filters and your eye. Does light get through?

_____

Step 8: Finally, sandwich the third filter at a 45° angle between the crossed pair. Does light get through?

_____

Summing Up

1. Why do polarized lenses make good sunglasses?

_____

2. Explain why the effects in Steps 1 to 3 occur.

_____

3. What parts of the sky are more partially polarized? What is your evidence?

_____

4. What evidence do you have that liquid crystal displays on calculators are or are not polarized?

_____

5. Explain your observations in Steps 6 to 8. If you're not into vectors, good luck!

_____

_____

Polaroids Ho!

CONCEPTUAL **Physical Science** **Experiment**

Atomic Size

Thickness of a BB Pancake

Purpose
To estimate the diameter of a BB.

Required Equipment and Supplies
75 mL of BB shot
100-mL graduated cylinder
tray
ruler
micrometer

Discussion
This activity distinguishes between *area* and *volume*, and sets the stage for the follow-up experiment *Oleic Acid Pancake*, where you will estimate the size of molecules. To see the difference between area and volume, consider eight wooden blocks arranged to form a single 2 x 2 x 2 inch cube. Since any cube has 6 sides, do you agree the outer surface area will be 6 times the area of one face — 24 square inches? The cube form exposes the minimum surface area (which is why buildings in cold areas are most often cubical in shape). Will not the surface area be more in any other configuration? For example, consider an arrangement of a 1 x 2 x 4 inch rectangular block — will not the outer surface area be greater? And if the blocks are spread out to form a stack only one cube thick, 1 x 1 x 8 inches, will not the outer surface area be maximum? Discuss these questions and answers with your lab partners.

The different configurations have different surface areas, but the volume remains constant. The volume of pancake batter is also the same whether it is in the mixing bowl or spread out on a surface (except that on a hot griddle the volume increases because of the expanding gas bubbles that form as the batter cooks). The volume of a pancake equals the surface area of one flat side multiplied by the thickness. If both the volume and the surface area are known, then the thickness can be calculated.

Volume = area x thickness

so simple rearrangement gives: Thickness = $\dfrac{\text{volume}}{\text{area}}$

Instead of cubical blocks or pancake batter, consider a graduated cylinder that contains BBs. The space taken up by the BBs is easily read as volume on the side of the cylinder. If the BBs are poured into a tray, their volume remains the same. Can you think of a way to estimate the diameter (or thickness) of a single BB without measuring the BB itself? Try it in this activity and see. It will be simply another step smaller to consider the size of molecules.

Thickness of a BB Pancake

Procedure

Step 1: Use a graduated cylinder to measure the volume of the BBs. (Note that 1 mL = 1 cm^3.)

Volume = _____ cm^3

Step 2: Carefully spread the BBs out to make a compact layer one pellet thick on the tray. With a ruler, determine the area covered by the BBs. Describe your procedure and show your computations.

Area = _____ cm^2

Step 3: Using the area and volume of the BBs, estimate the diameter of a BB. Show your computations.

Estimated diameter = _____ cm

Step 4: Check your estimate by using a micrometer to measure the diameter of a BB.

Measured diameter = _____ cm

Summing Up

1. What assumptions did you make when estimating the diameter of the BB?

_____

2. How does your estimate and the measurement of the diameter of the BB compare? Calculate the percentage difference (consult the Appendix on how to do this) between the measured and estimated diameter of the BB.

_____

_____

3. Oleic acid is an organic substance that is soluble in alcohol but insoluble in water. When a drop of oleic acid is placed in water, it usually spreads out over the water surface to create a *monolayer*, a layer that is one molecule thick. From your experience with BBs, describe a method for estimating the size of an oleic acid molecule.

_____

_____

_____

Thickness of a BB Pancake

CONCEPTUAL **Physical Science** ━━━━━━━━━━━━ **Experiment**

Molecular Size

Oleic Acid Pancake

Purpose
To estimate the size of a molecule of oleic acid.

Required Equipment and Supplies
tray
water
chalk dust or lycopodium powder
eye dropper
oleic acid solution (5 mL oleic acid in 995 mL of ethanol)
10-mL graduated cylinder

Discussion
During this experiment you will estimate the *diameter* of a single molecule of oleic acid! The procedure for measuring the diameter of a molecule will be much the same as that of measuring the diameter of a BB in the previous activity. The diameter is calculated by dividing the volume of the drop of oleic acid used, by the area of the *monolayer* film that is formed. The diameter of the molecule is the depth of the monolayer.

DID YOU EVER THINK YOU COULD ESTIMATE THE SIZE OF A SINGLE MOLECULE?

PHYSICS! ≈SIGH≈

Volume = area x depth

$$\text{Depth} = \frac{\text{volume}}{\text{area}}$$

Procedure
Step 1: Pour water into the tray to a depth of about 1 cm. So that the acid film will show itself, spread chalk dust or lycopodium power very lightly over the surface of the water; too much will "hem in" the oleic acid.

Step 2: Using the eye dropper, gently add a single drop of the oleic acid solution to the surface of the water. When the drop touches the water, the alcohol in it will dissolve in the water, but the oleic acid will not. The oleic acid spreads out to form a nearly circular patch on the water. Measure the diameter of the oleic acid patch in several places, and compute the average diameter of the circular patch.

Average diameter = _____ cm

The average radius is, of course, half the average diameter. Now compute the area of the circle ($A = \pi r^2$).

Area of circle = _____ cm^2

Step 3: Count the number of drops of solution needed to occupy 1 mL (or 1 cm^3) in the graduated cylinder. Do this three times, and find the average number of drops in 1 cm^3 of solution.

Oleic Acid Pancake

Number of drops in 1 cm^3 = _____

Divide 1 cm^3 by the number of drops in 1 cm^3 to determine the volume of a single drop.

Volume of single drop = _____ cm^3

Step 4: The volume of the oleic acid alone in the circular film is much less than the volume of a single drop of the solution. The concentration of oleic acid in the solution is 5 mL per liter of solution. Every cubic centimeter of the solution thus contains only $\frac{5}{1000}$ cm^3, or 0.005 cm^3, of oleic acid. The volume of oleic acid in one drop is thus 0.005 of the volume of one drop. Multiply the volume of a drop by 0.005 to find the volume of oleic acid in the drop. This is the volume of the layer of acid in the tray.

Volume of oleic acid = _____ cm^3

Step 5: Estimate the diameter of an oleic acid molecule by dividing the volume of oleic acid by the area of the circle.

Diameter = _____ cm

The diameter of an oleic acid molecule as obtained by this method is good, but not precise. This is because an oleic acid molecule is not spherical, but rather elongated like a hot dog. One end is attracted to water, and the other end points away from the water surface. The molecules stand up like people in a puddle! So the estimated diameter is actually the estimated length of the short side of an oleic acid molecule.

Summing Up
1. What is meant by a *monolayer*?

_____

2. Why is it necessary to dilute the oleic acid for this experiment? Why alcohol?

_____

3. The shape of oleic acid molecules is more like that of a hot dog than a sphere. Furthermore, one end is attracted to water (*hydrophyllic*) so that the molecule stands up on the surface of water. Assume an oleic molecule is 10 times longer than it is wide. Then estimate the volume of one oleic acid molecule.

_____

_____

Oleic Acid Pancake

CONCEPTUAL **Physical Science** ————————————————

Chain reaction

Chain Reaction

Purpose
To simulate a simple chain reaction

Required Equipment and Supplies
100 dominoes
large table or floor space
stop watch

Discussion
Give your cold to two people who in turn give it to two others
who in turn do the same on down the line and before you know
it everyone in class is sneezing. You have set off a chain
reaction. Similarly, when one electron in a photomultiplier
tube in certain electronic instruments hits a target that
releases two electrons that in turn do the same on down the
line, a tiny input produces a large output. When one neutron
triggers the release of two or more neutrons in apiece of
uranium, and the triggered neutrons trigger others in
succession, the results can be devastating. In this activity
we'll explore this idea.

Fig. A

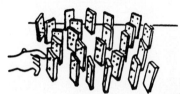

Procedure
Step 1: Set up a strand of dominoes about half a domino length apart in a straight line.
Gently push the first domino over, and measure how long it takes for the entire strand to
fall over (just like those television commercials).

Step 2: Arrange the dominoes as in Figure A, so that when one domino falls, another one or
two are toppled over. These topple others in chain reaction fashion. Set up until you run
out of dominoes or table space. When you finish, push the first domino over and watch the
reaction. Notice the number of the falling dominoes per second at the beginning versus the
end.

Summing Up
1. Which reaction, wide-spaced or close-spaced dominoes took a shorter time?

_____

_____

2. How did the number of dominoes being knocked over per second change for each
reaction?

_____

_____

_____

Chain Reaction

3. What caused each reaction to stop?

_____

_____

4. Now imagine that the dominoes are the neutrons released by uranium atoms when they fission (split apart). Neutrons from the nucleus of a fissioning uranium atom hit other uranium atoms and cause them to fission. This reaction continues to grow if there are no controls. Such an uncontrolled reaction occurs in a split second and is called a *nuclear explosion*. How is the domino chain reaction similar to the nuclear fission process?

_____

_____

_____

5. How is the domino reaction dissimilar to the nuclear fission process?

_____

_____

Chain Reaction

CONCEPTUAL **Physical Science** ———————————————— **Experiment**

Atomic Emissions

Bright Lights

Purpose
To examine the light emitted by various elements.

Required Equipment and Supplies
Bunsen burner
metal spatulas or wire
dilute HCl solution (0.1 M)
metal salts (powdered): lithium chloride, sodium chloride, potassium chloride,
      calcium chloride, strontium chloride, barium chloride, copper chloride
spectral tubes containing various gases
diffraction gratings
spectroscope (commercial or homemade from a cardboard tube and aluminum foil)

Discussion
When materials are heated their atoms emit light. The color of the light is characteristic
of the type of atoms in the heated material. Lithium atoms, for example, glow scarlet red,
and copper atoms glow leprechaun green. An atom emits light when its electrons make a
transition from higher energy levels to lower ones. Every element has its own characteristic
pattern of electron energy levels and therefore emits its own characteristic pattern of light
frequencies (colors) when excited (see Chapter 14, *Conceptual Physical Science*).

It is interesting to look at the light emitted by ionized and gaseous elements through a
diffraction grating or prism. Rather than producing a full continuous spectrum of colors, these
elements produce a spectrum that is *discontinuous* showing only particular colors
(frequencies). When the light passes through a thin slit, the different colors appear as a
series of vertical lines (Figure 1). Each vertical line corresponds to a particular energy
transition for an electron. The pattern of lines is characteristic of the element and is often
used as an identifying feature–much like a fingerprint. Astronomers can tell the elemental
composition of far-away stars by closely examining their spectra.

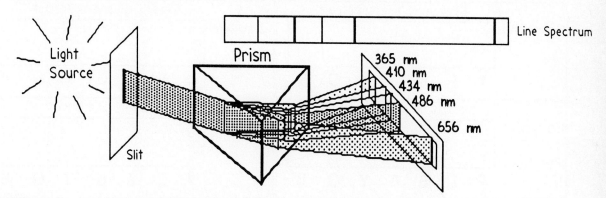

Figure 1. Ionized or gaseous elements when heated produce a discontinuous spectrum of light.

Bright Lights

Safety

Remove all combustible materials from your work station when working with a Bunsen burner. Be careful not to touch the tip of the metal spatula while it is hot.

Procedure

Flame Test

Step 1: Clean the tip of a metal spatula by holding it in a Bunsen burner flame until it is red hot, and then dip it into a dilute solution of HCl. Repeat this process several times until you no longer see color coming from the metal when heated.

Step 2: Obtain small amounts of the metal salts to be tested. Label each sample. Dip the spatula in the HCl solution and then directly into one of the salts. Put the tip of the spatula with adhering crystals into the flame and observe the color. Record your observations in Table 1. Repeat this procedure for all the salts. Rinse in water and then clean the spatula as in Step 1 between each testing.

Table 1.

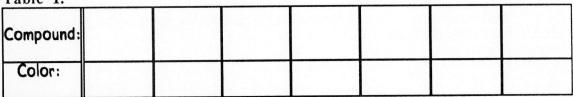

Compound:							
Color:							

Step 3: Repeat Step 2, this time observing the flame through a spectroscope, which consists of a tube or box containing a narrow slit at one end and a diffraction grating at the other. Your spectroscope may be easily built using a cardboard tube, aluminum foil, and a diffraction grating (Figure 2). Mount the spectroscope on a ring stand. Sketch the predominate lines you observe for each element in Table 2. Use color pencils if you have them. You will see lines both to the left and the right of the slit. Sketch only the lines toward the right hand side. (Note: Some elements will also show regions of continous color.)

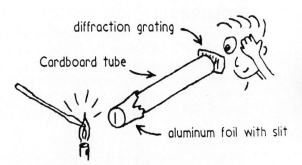

Figure 2. A simple spectroscope.

Step 4: Obtain an unknown metal salt from your instructor. Identify it based upon its line spectrum, which you should sketch in the last cell of Table 2 (next page).

Table 2. Metal Salts

Compound	Sketch of line spectrum
	V I B G Y O R
Compound	Sketch of line spectrum
	V I B G Y O R
Compound	Sketch of line spectrum
	V I B G Y O R
Compound	Sketch of line spectrum
	V I B G Y O R

(continued...)

Table 2. Metal Salts (cont.)

Compound	Sketch of line spectrum
	V I B G Y O R

Compound	Sketch of line spectrum
	V I B G Y O R

Compound	Sketch of line spectrum
	V I B G Y O R

Compound Unknown	Sketch of line spectrum
	V I B G Y O R

Gas Discharge Tubes

Step 5: Observe through a diffraction grating the light emanating from various gas discharge tubes. You need not pass the light through a slit because the discharge tubes themselves are narrow. Observe and sketch hydrogen, oxygen, and water gases if available — then answer question #7 on the next page.

Table 2. Gases

Compound	Sketch of line spectrum
	V I B G Y O R

Compound	Sketch of line spectrum
	V I B G Y O R

Compound	Sketch of line spectrum
	V I B G Y O R

Compound	Sketch of line spectrum
	V I B G Y O R

Compound	Sketch of line spectrum
	V I B G Y O R

Compound	Sketch of line spectrum
	V I B G Y O R

Compound	Sketch of line spectrum
	V I B G Y O R

Compound	Sketch of line spectrum
	V I B G Y O R

Compound	Sketch of line spectrum
	V I B G Y O R

Compound	Sketch of line spectrum
	V I B G Y O R

Bright Lights

Summing Up

1. What is the identity of your unknown metal salt?

<u>Unknown Number</u>   <u>Identity</u>

2. When the spatula was initially being cleaned in the flame it may have given off a yellow color. What residue was likely on the spatula before it was cleaned?

_____

_____

3. Suggest one reason why campfires are generally yellow instead of blue like the flame of the Bunsen burner?

_____

_____

4. What produces the color of fireworks?

_____

_____

5. What function does the narrow slit in the spectroscope serve? How do the spectra from the flame tests look without it?

_____

_____

6. Is the gas in a blue "neon" lamp actually neon? Briefly explain.

_____

_____

7. Does the line spectrum of water bear any resemblence to the line spectra of hydrogen and of oxygen? Why or why not?

_____

_____

Bright Lights

CONCEPTUAL **Physical Science** | **Experiment**

Nuclear Model of the Atom

Nuclear Pennies

Purpose
To measure the diameters of various beakers indirectly using a technique similar to the one Ernest Rutherford used to determine the size of the atomic nucleus.

Required Equipment and Supplies
a large number of pennies (20-50 per lab group)
a variety of beakers (10 mL, 25 mL, 50 mL, 100 mL)

Discussion
People sometimes have to resort to something besides a sense of sight to determine the shapes and sizes of things, especially for things smaller than the wavelength of light. One way to do this is to shoot particles at the object to be investigated, and study the deflected paths these particles take after interacting with the object. Ernest Rutherford inferred the size of the nuclei of gold atoms by studying how alpha particles were deflected by the nuclei in gold foil (Chapter 14, *Conceptual Physical Science*). In this experiment, you will infer the diameter of a beaker by shooting pennies at it.

You will slide pennies (representing alpha particles) lying flat on a smooth surface toward a set of beakers (representing nuclei), all of the same size and determine the diameters of the beakers from the ratio of hits to the total number of shots. It's a bit like throwing rocks at a line of trees while blindfolded. If you have very few hits per certain number of throws then the trees "feel" small.

Statistical Analysis: Formula for Calculating Diameter of Beaker
When you shoot a penny toward a lone beaker, there is a certain probability of a hit between the sliding penny and the beaker. One expression of the probability P of a hit is the ratio of the target width to the path width W (see Figure 1).

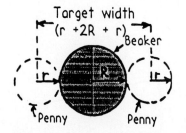

$$P = \frac{\text{target width}}{\text{path width}}$$

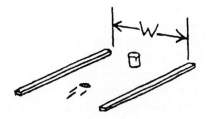

Figure 1. The probability of a sliding penny colliding with a lone beaker is directly proportional to the target width and inversely proportional to the path width W.

The target width is equal to the radii of two pennies plus the diameter of the beaker, as shown in Figure 1. The probability that a penny will hit a lone beaker within the path width, therefore, is

$$P = \frac{\text{target width}}{\text{path width}} = \frac{(2R + 2r)}{W} = \frac{2(R + r)}{W}$$

where R = the unknown radius of the beaker
  r = the radius of sliding penny
 $R + r$ = the distance between the centers of the penny and the beaker when touching
 and W = the path width

If the number of beakers is increased to a value of N, the probability of a hit is increased by a factor of N. Thus, the probability that a sliding penny will hit one of the N beakers is

$$P = \frac{2N(R + r)}{W}$$

The probability of a hit can also be determined experimentally by the ratio of the number of hits to the total number of shots.

$$P = \frac{H}{S}$$

where H = the number of hits
and S = the number of shots

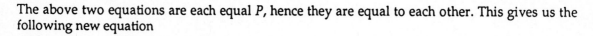

For one target marble,
$$p = \frac{\text{target width}}{\text{path width}} = \frac{2(R + r)}{W}$$

But probability by experiment equals the number of hits per number of shots!
$$p = \frac{H}{S}$$

The above two equations are each equal P, hence they are equal to each other. This gives us the following new equation

$$\frac{2N(R + r)}{W} = \frac{H}{S}$$

Algebraically, this means that

$$R + r = \frac{HW}{2NS}$$

Recall that R equals the radius of the beaker, which need not be measured directly. All other quantities can be determined experimentally and then plugged into this equation to solve for R. Multiply by 2 to find the beaker's diameter.

Procedure

Step 1: Work in groups of 2 or 3. Elect one student to be the penny shooter and have him or her leave the room temporarily. Lay down two meter sticks parallel to each other 60 cm apart (W = 60 cm) on a flat smooth surface. Secure them to the surface by taping down the ends. Choose a particular size beaker and fill 6 to 9 of these beakers half-way with water. Place the beakers in between the meter sticks toward one side as in Figure 2. Distribute the beakers evenly and place a cardboard box with both the top and bottom cut out over the beakers and resting on the meter sticks. The box will serve to hide the beakers from the penny shooter.

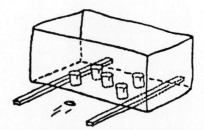

Figure 2.

Nuclear Pennies

Step 2: Bring the penny shooter back to the lab station and have him or her slide pennies flat on their sides one at a time under the box toward the beakers. The pennies should be sent fast enough so they can pass through if unimpeded. Also, the pennies need to be sent as randomly as possible. A good way to do this is to knock each penny with a second penny that you hold under your index finger. Switching between right and left hands also helps to achieve randomness. If the same penny hits two nuclear beakers it counts as just one hit. Also, it does not matter if the sliding penny richochets off of a meter stick. A significant number of shots needs to be made — at least 200 — before the results become statistically significant. Remove pennies as they get stopped within the area of the beakers to avoid pennies richocheting off pennies. Record your total number of hits, H, and total number of shots, S, in Table 1.

Step 3: Measure the radius of a penny and use your experimental data to predict the diameter of each of the beakers (Table 1). Allow the penny shooter to inspect an assortment of beakers and challenge him or her to figure what size beakers were hidden in the box.

Step 4: Calculate your percent error using the following equation

$$\text{Percent Error} = \frac{\text{difference between probable and actual diameters}}{\text{actual diameter}} \times 100$$

Step 5: For a second trial, elect a second penny shooter and repeat the above procedures using a different sized beaker. Perform additional trials so that every one in the group has a chance to be the penny shooter.

Going Further

Step 6: Play around with other variables besides the size of the beaker. For example, perform a trial where you change the path width, W.

Table 1.

	Trial 1	Trial 2	Trial 3	Trial 4
W (path width)				
N (number of beakers)				
H (number of hits)				
S (number of shots)				
r (radius of penny)				
R (calculated radius of beaker)				
Calculated Diameter of Beaker				
Actual Diameter of Beaker (using ruler)				
Percent Error				

Nuclear Pennies

Summing Up

1. What is the probability of a hit when the target width is just as large as the path width?

_____

_____

2. Would your calculated diameter of the beaker likely be greater or smaller than the actual diameter if the beakers were not distributed evenly but lined up in a single row parallel to the meter sticks?

_____

_____

3. Suggest a reason for having water in the beakers.

_____

_____

4. State a conclusion you can draw from this experiment.

_____

_____

Nuclear Pennies

CONCEPTUAL **Physical Science**   **Experiment**
Radioactive Half Life

To Half or Half Not

Purpose
To simulate radioactive decay half life.

Required Equipment and Supplies
25 small color-marked cubes per group (one side red, two sides blue, three sides blank). Spray-painted sugar cubes work well. Multifaceted dice may also be used.

Discussion
The rate of decay for a radioactive isotope is measured in terms of **half life** — the time for one half of a radioactive quantity to decay. Each radioactive isotope has its own characteristic half life (Table 1). For example, the naturally occurring isotope of uranium, uranium-238, decays into thorium-234 with a half life of 4.51×10^9 years. This means that only half of an original amount of ^{238}U remains after this time. After another 4.51×10^9 years half of this decays leaving only one fourth of the original amount remaining. Compare this with the decay of polonium-214, which has a half life of 1.6×10^{-4} seconds. With such a short half life, any sample of polonium-214 will quickly disintegrate.

Table 1

Element	Half Life
Uranium-238	4.51×10^9 years
Plutonium-239	2.44×10^4 years
Carbon-14	5.73×10^3 years
Lead-210	20.4 years
Bismuth-210	5.0 days
Polonium-214	1.6×10^{-4} seconds

The half life of an isotope can be calculated by the amount of radiation coming from a known quantity. In general, the shorter the half life of a substance, the faster it decays, and the more radioactivity per amount is detected.

In this activity, you will investigate three hypothetical substances, each represented by a color on the face of a cube. The first substance, represented by a given color, is marked on only one side of the cube. The second substance, represented by a second color, is marked on two sides of the cube, and the third substance, represented by a third color (or lack thereof), is marked on the remaining three sides. The process of decay for these substances is simulated by rolling a large number of these identically painted cubes. As a substance's color turns face up, it is considered to have decayed and is removed from the pile. This process is repeated until all of the cubes have been removed. Since the color of the first substance is only on one side, this substance will decay the slowest, that is, its color will fall face up least frequently and it will take the longest before all the cubes are removed. The second substance, marked on two sides, will decay faster requiring fewer rolls before all the cubes are removed. The third substance, marked on three sides, will decay the fastest. After tabulating and graphing the numbers of cubes that decay in each roll for these simulated substances, you will be able to determine their half-lives.

To Half or Half Not

Procedure

Step 1: Shake the cubes in a container and roll them onto a flat surface.

Step 2: Count the red faces that are up and record this number under "Removed" in the Data Table.

Step 3: Remove the red cubes in a pile off to the side.

Step 4: Gather the remaining cubes back into the container and roll them again.

Step 5: Repeat steps 2-4 until all cubes have been counted, tabulated and set aside.

Step 6: Repeat steps 1-5 removing cubes that show the blue faces up.

Step 7: Repeat steps 1-5 removing cubes that show the white faces up.

Data Table

Throw	Red		Blue		White	
	Removed	Remaining	Removed	Remaining	Removed	Remaining
Initial Count						
1						
2						
3						
4						
5						
6						
7						
8						
9						
10						
11						
12						
13						
14						
15						
16						
17						
18						
19						
20						
21						
22						
23						
24						
25						

To Half or Half Not

Step 8: Plot the number of cubes remaining verses the number of throws for each substance on the following graph. Use a different color to graph the results for each substance. For each color, draw a single smooth line or curve that approximately connects all points. DO NOT CONNECT THE DOTS! Provide a color code.

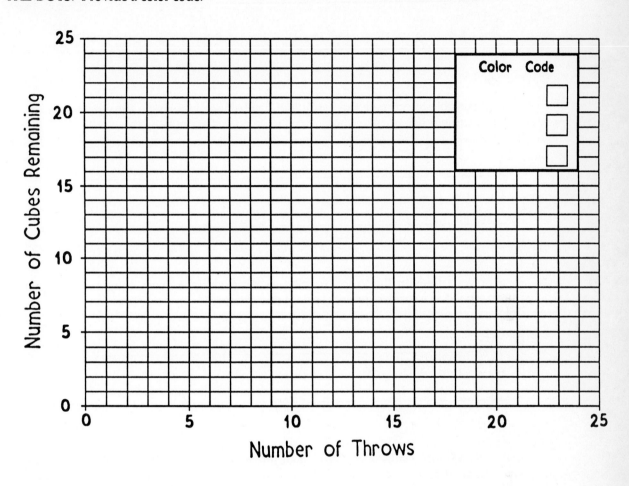

Summing Up

1. How many rolls did it take for the number of each colored cube to be reduced by half? These are your half life readings.

   Red _____   Blue _____   White _____

2. The half life of a decaying substance is measured in units of time. What is the unit of half life used in this simulation?

   _____

3. In each case, how many rolls did it take to remove all of the cubes?

   Red _____   Blue _____   White _____

4. Which of these hypothetical substances would be the most harmful?

   _____

To Half or Half Not

5. How might you simulate the radioactive decay of a substance that decays into a second substance that also decays?

_____

_____

6. Is it possible to estimate the half life of a substance in a single throw? How accurate might this estimate be?

_____

_____

7. Are your lines in the graph for Step 8 fairly straight or do they curve? Do these lines correspond to a constant or nonconstant rate of decay?

_____

_____

8. a. 1000 grams of substance X has a half life of 10 years, how much will be left after:
   - 10 years?

   - 20 years? _____

   - 50 years? _____

   - 100 years? _____

   b. Will this sample of substance X ever totally disappear? If so, estimate how soon. If not, explain.

   _____

   _____

To Half or Half Not

CONCEPTUAL **Physical Science**
Percent Oxygen in Air

Tubular Rust

Purpose
To determine the percent oxygen in air.

Required Equipment and Supplies
steel wool
test tube (15 cm)
glass rod
beaker
metric ruler and rubber band
white distilled vinegar (5% acetic acid)

Discussion
This activity takes advantage of the rusting of iron by oxygen to determine the percent oxygen in the air. Iron is placed in an air-filled test tube, which is then inverted in water. As the iron reacts with the oxygen the pressure inside the test tube decreases, and atmospheric pressure pushes water up into the tube. The decrease in volume of air in the test tube corresponds to the depletion of its oxygen content. By measuring the quantity of air in the test tube before and after the rusting of the iron the percent oxygen in air by volume can be calculated.

Procedure
Step 1: Measure about 1 gram of steel wool that has been pulled apart to increase its surface area. To promote rusting, dip the wool into some white distilled vinegar. Shake off the excess vinegar, and push the steel wool half-way into a test tube 15-cm in length using a glass rod. The wool should be packed as loosely as possible to maximize its surface area but packed tight enough so that it doesn't fall out upon inverting the tube.

Step 2: Attach a 15-cm ruler to the test tube using a rubber band. Position the ruler so that the *0 mm* mark is toward the open end. Carefully invert the test tube into a beaker of water. Clamp the test tube to a ring stand and adjust its height so that the lip is just under the water. Adjust the ruler so that the *0 mm* mark is even with the water level inside the tube.

Step 3: Water will climb up into the test tube as the reaction proceeds. As this occurs lower the test tube deeper into the beaker so that the water levels outside and inside the test tube remain even. Begin to read and record the water level inside the test tube at 3 minute intervals until it stops rising.

Step 4: Plot a graph of the water level inside the test tube verses time.

Step 5: Calculate the fraction of oxygen in air as the maximum height of the water inside the tube divided by the total length of the tube. Mulitply by 100 to obtain a percentage.

Tubular Rust

Summing Up

1. What is your experimental percent oxygen in the air?

_____

2. Why is it important that the water levels inside and outside the test tube remain even?

_____

_____

_____

_____

3. Would it take a longer or shorter time for the oxygen to be depleted if the steel wool were packed tightly at the bottom of the test tube. Briefly explain.

_____

_____

_____

4. Your calculations for finding the percent oxygen in air neglect the volume occupied by the steel wool. Would taking the steel wool's volume into account increase or decrease your calculated percent oxygen? Briefly explain.

_____

_____

_____

_____

Tubular Rust

CONCEPTUAL **Physical Science**

Collection of a Gas

Collecting Bubbles

Purpose
To isolate gaseous carbon dioxide by water displacement.

Required Equipment and Supplies
baking soda
vinegar
Erylenmeyer flasks (2 x 250 mL)
beaker (1000 mL)
rubber stoppers (2, one with glass rod inserted through it)
tubing with a paper clip inserted in one end
ring stand with clamp

Discussion
A common method of collecting a gas produced from a chemical reaction is by the displacement of water. Bubbles of the gas are directed into an inverted flask filled with water. As the gas rises into the container it displaces water. Once all the water is displaced the flask may be sealed with a stopper. The physical and chemical properties of the gas can then be investigated. In this activity baking soda (sodium bicarbonate, $NaHCO_3$), and vinegar (5% acetic acid, CH_3COOH), will be reacted to form gaseous carbon dioxide, CO_2, which will be collected by water displacement. The other products of this reaction, water, H_2O, and sodium acetate, $CH_3COO^-Na^+$, form a liquid phase that remains in the reaction vessel.

$$NaHCO_3 \ + \ CH_3COOH \ \longrightarrow \ CO_2 \ + \ H_2O \ + \ CH_3COO^- \ Na^+$$

sodium bicarbonate   acetic acid   carbon dioxide   water   sodium acetate

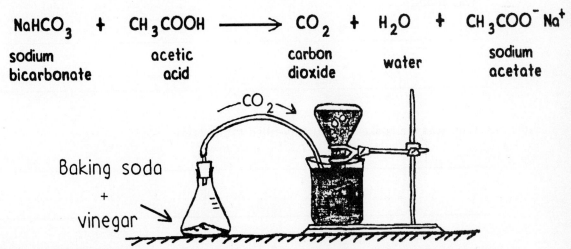

Figure 1. Baking soda (sodium bicarbonate) and vinegar (acetic acid) react to form gaseous carbon dioxide, CO_2.

Procedure
Step 1: Assemble the set-up as shown in Figure 1. A 250 mL Erlenmeyer flask is equipped with a rubber stopper through which a glass tube has been inserted. The glass tube is attached to a plastic tube that is forced into a J-shape on the opposite end by the bending of an inserted paper clip. A second 250 mL Erlenmeyer flask is filled with water and inverted

Collecting Bubbles

into a 1000 mL beaker filled with 700 mL of water. The J-shaped end of the plastic tubing is then fixed below the lip of the inverted flask.

Step 2: Unstopper the upright Erlenmeyer flask and add about a tablespoon of baking soda (sodium bicarbonate) followed by a capful of vinegar (5% acetic acid). Allow the vigorous bubbling to settle down. Add a second capful of vinegar and immediately stopper the upright flask. This will cause bubbles to fill the inverted flask.When bubbling stops add an additional capful of vinegar to collect additional bubbles (carbon dioxide). Continue in this manner until the inverted beaker has been filled with carbon dioxide. *(Caution: Add vinegar only by the capful. Froth may overflow the flask if you add too much vinegar all at once.)*

Step 3: Stopper the inverted flask tightly to seal the carbon dioxide.

Step 4: Light a small candle and experiment with your newly isolated gas.

Going Further
Step 5: Measure the mass of the carbon dioxide you collected using a balance that measures to the nearest milligram. Note: the density of air at sea level and room temperature is about 1.17 g/L.

Summing Up
1. Is carbon dioxide more or less dense than air? How do you know?

_____

_____

2. Why was the flask containing baking soda not stoppered after the first capful of vinegar was added?

_____

_____

3. Why was the beaker not initially filled to the brim?

_____

_____

Collecting Bubbles

CONCEPTUAL **Physical Science**

Growth of Large Crystals

Crystal Clear

Purpose
To prepare a large ionic crystal.

Required Equipment and Supplies
beaker
nylon thread
fast-drying epoxy glue
wire
ionic compounds
     sodium chloride, $NaCl$
     nickel sulfate hexahydrate, $NiSO_4 \cdot 6H_2O$
     nickel sulfate heptahydrate, $NiSO_4 \cdot 7H_2O$
     potassium sulfate, K_2SO_4
     copper sulfate pentahydrate, $CuSO_4 \cdot 5H_2O$
     copper acetate monohydrate, $Cu(CH_3COO)_2 \cdot H_2O$
     calcium carbonate, $CaCO_3$
     sodium nitrate, $NaNO_3$

Discussion
Crystals of different compounds have different shapes because of the variety of ways atoms are able to pack. With sodium chloride, $NaCl$, for example, sodium and chlorine ions pack in a cubic form. This gives rise to cubically shaped sodium chloride crystals. With calcium carbonate, $CaCO_3$, on the other hand, calcium and carbonate ions pack together in a rhombic orientation. This gives rise to rhombohedrally shaped calcium carbonate crystals (Figure 1).

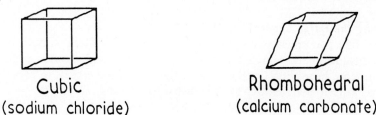

Cubic
(sodium chloride)

Rhombohedral
(calcium carbonate)

Figure 1. The macroscopic shape of a crystal is a consequence of the microscopic arrangement of atoms.

Ionic crystals are fairly easy to prepare. One method involves dissolving the chosen ionic compound (salt) in a solvent to the point of saturation at a temperature above room temperature. As the solution cools the solubility of the solute decreases and some of the solute comes out of solution (See Chapter 18, *Conceptual Physical Science*). If this process happens slow enough the solute ions pack together in a crystalline form. As more and more ions pack together the crystal gets larger — it grows. Crystals can also be prepared from a saturated solution without altering the temperature. Over time the solvent evaporates. This causes some of the solute to come out of solution, which permits the formation of crystals.

Crystal Clear

Small crystals can form within an hour using the method of cooling a saturated solution. Larger crystals — on the order of a few centimeters — are best formed by allowing the solvent to evaporate slowly. In practice, the best way to form a large crystal is to immerse a small crystal in a saturated solution. Solute will preferentially crystallize on the surface of the small crystal. Within a matter of weeks, the small crystal grows into a large one.

Procedure

Obtaining Small Crystals
Step 1: Select an ionic compound and prepare about 200 mL of a saturated solution at room temperature. Use ample stirring to make sure your solution is saturated.

Step 2: Add a small excess of the compound to the solution after it is saturated. Elevate the temperature of the solution so that this added quantity dissolves. Cover the beaker with a watch glass and allow the solution to cool slowly undisturbed. If no crystals form, try scratching the bottom of the beaker with a metal spatula. Be patient. If crystals still don't form, this indicates that the solution is not saturated and you still need to complete Step 1.

Step 3: Filter the solution and collect the better shaped crystals. Do not throw away the solution. It will be used in subsequent steps.

Growing a Large Crystal
Step 4: Pour the filtered solution into a clean beaker. Saturate the solution once again with your selected compound. You may or may not want to play with the temperature in resaturating this solution.

Step 5: Fix the best shaped crystal from Step 3 to a nylon thread using fast-drying epoxy glue. Tie the opposite end of the thread to a wire that can lay across the mouth of the beaker, and lower the crystal midway into the saturated solution as in Figure 2. Carefully set the beaker aside in a place where it may remain undisturbed for several weeks.

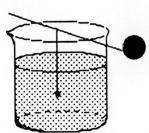

Step 6: Check up on your growing crystal periodically. During spare moments of future laboratories you may wish to alter the conditions or modify your procedures to maximize crystal growth. Be creative. Use your own judgment and scientific sense.

Figure 2.

Collecting the Large Crystal
Step 7: When you are satisfied with the size of your crystal it should be removed from the saturated solution and allowed to dry. Your dried crystal should be stored in a sealed jar to protect it from humidity. On the jar note the size of the crystal in centimeters and the mass in grams. Turn in your sample to your instructor.

Summing Up
1. How can you tell when a solution is saturated?

_____

2. What is the difference between a crystal and an amorphous solid such as glass?

_____

_____

Crystal Clear

CONCEPTUAL **Physical Science**
Radial Paper Chromatography

Circular Rainbows

Purpose
To separate the different colored components of black ink.

Required Equipment and Supplies
a variety of black felt-tip pens
circular filter paper or chromatography paper
a variety of solvents
> water
> methanol (wood alcohol)
> ethanol (grain alcohol)
> isopropyl alcohol (rubbing alcohol)
> white distilled vinegar
> acetone (finger nail polish remover)
beakers or crucibles upon which to place the paper

Discussion
What makes black ink black? Black ink is made by combining many different colored inks such as blue, red, and yellow. Together, these inks serve to absorb all the frequencies of light. With no light reflected, the ink appears black.

Procedure
It's easy to separate the components of an ink using a chemical technique called "paper chromatography." Place a concentrated dot of the ink at the center of a porous piece of paper, such as filter paper. Prop the marked paper on a small beaker or crucible, then carefully add a drop of solvent such as water, white vinegar, or rubbing alcohol on top of the dot. Watch the ink spread radially with the solvent. Since the various components have differing affinities for the solvent, we find that they travel with the solvent at different rates. Just before your drop of solvent is completely absorbed, add a second drop, then a third, and so on until the components have separated to your saticefaction. How the components separate depends on several factors, including your choice of solvent and your technique. Black felt-tip pens tend to work best, but you should experiment with a variety of different types of pens, or even food coloring. It's also interesting to watch the leading edge of the moving ink under a microscope. Check for capillary action!

Summing Up
1. The different components of black ink not only have differing affinities for the solvent, but have differing affinities for the paper, which is fairly polar. How might you expect an ionically-charged component to behave while using a relatively nonpolar solvent, such as acetone — would it readily travel with the solvent or might it stay behind?

_____

2. Have you noticed that all "blue" inks aren't the same color blue? How might this be explained?

_____

Circular Rainbows

CONCEPTUAL **Physical Science** ⌐ Activity ⌐

Purification of Brown Sugar

Home Brew Sugar

Purpose
To isolate white sugar from brown sugar.

Required Equipment and Supplies
Brown sugar; Empty ketchup bottle; Kitchen knife or microspatula;
Small wide mouth jar (such as a baby food jar)

Discussion and Procedure
A supersaturated solution of brown sugar in water will be prepared by dissolving ample amounts of brown sugar in a small amount of boiling water (1/2 cup). The brown sugar will be added up to the point that major frothing occurs. The thick syrupy solution will be allowed to cool and then poured into an empty ketchup bottle, which will be allowed to stand still for many days until a significant number of crystals have formed. After crystal growth, the syrupy solution, hereafter called "mother liquor," will be poured out by inverting the ketchup bottle into a second container (drainage may take several hours). The similarity of the mother liquor to molasses will be noted. The crystals that remain behind in the ketchup bottle will be rinsed once with warm water to remove additional mother liquor. A few of these crystals will be collected using a knife and their properties examined. Remaining crystals will be dissolved in small amounts of hot water and collected in a small cooking pot. This solution may be concentrated by boiling away any excess water. After cooling, the solution will be poured into a small wide mouth jar and allowed to remain still for many days until significant numbers of crystals have formed. After crystal growth, the mother liquor of this solution will be poured out. The crystals that remain may be collected onto a towel, rinsed briefly with warm water and quickly dried so that they retain crystal shape. If crystals still retain significant brown color, they may be recrystallized using the general procedures given above.

Summing Up
1. How might you prove that commercial grade white sugar contains small amounts of molasses?

_____

2. Which is more "pure," white sugar or brown sugar? (circle one)

3. Which is more "natural", white sugar or brown sugar? (circle one)

4. On a separate sheet of paper (or back side of this) write a statement regarding the quality verses quantity of the sugar crystals you obtained from this activity.

Further Research (optional)
Insects just love sugar solutions, but do they love sugar when it's pure and dry? Are they attracted to a concentrated solution of Nutrasweet artificial sweetener?

Name _____   Section _____ Date _____

CONCEPTUAL **Physical Science** _____ | **Activity** |

Polar Interactions

Polarity of Molecules

Purpose
To investigate the effect of a charged rod on tiny streams of liquids.

Required Equipment and Supplies
sewing needle
buret and support
water
triclorotriflouroethane (TTE)

Discussion
A water molecule consists of one atom of oxygen and two atoms of
hydrogen. The water molecules themselves have no net electrical
charge, but there is a slight displacement of positive and negative
charges in their arrangement. This is because the oxygen nucleus has
a greater electric charge than either hydrogen nucleus, and the
electrons of the two hydrogen atoms are more attracted to the oxygen
nucleus and spend more time near it than either of the two hydrogen
atoms. The water molecule is electrically *polarized* — where opposite charges on opposite ends
create an electric *dipole*. The oxygen end of the molecule behaves as if it were slightly
negatively charged and the hydrogen end behaves as if it were slightly positively charged.

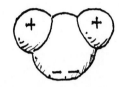

The polarity of water molecules accounts for some of water's unique properties. The positive
end of one water molecule is attracted to the negative end of another to form weak bonds. In the
solid state, the atoms group in such a way as to create a hexagonal crystal lattice that occupies
more space than the individual unbonded atoms. So ice floats. When the temperature increases
and the ice melts to form water, the weak attraction between water molecules causes *surface
tension*, making it possible to float needles and razor blades. As the temperature increases, the
molecules are not so able to form bonds and surface tension decreases.

Procedure
Step 1: Carefully float a needle on the surface of water at room temperature. Heat the water
and see what happens. Explain your observations.

_____

_____

Step 2: Briskly rub a piece of cat's fur on a hard rubber rod to charge the rod
negatively. Bring the charged rod near a stream of water from a buret
prepared by your instructor. What happens to the stream of water? Why?

buret

charged rod

← stream of water

_____

_____

Polarity of Molecules

Step 3: What do you think would happen if you were to use a positively charged rod? Would the stream behave differently than before? Try it and report what you see.

_____

_____

Step 4: Repeat using a non-polar substance such as triclorotrifluoroethane (TTE). How do your results compare?

_____

_____

Summing Up
Make a sketch of how the charges in the falling water arrange themselves in the presence of a negatively charged rod. Make another sketch for a positively charged rod.

Polarity of Molecules

CONCEPTUAL **Physical Science**

Synthesis of Nylon

A Roping Experience

Purpose
To make a continuous thread of nylon several meters long by mixing two immiscible solutions in a small beaker.

Required Equipment and Supplies
50 mL Beaker
bent paper clip
disposable gloves
Solution A
  6% hexamethylenediamine (1,6-diaminohexane) in 0.5 M NaOH
Solution B
  5% adipoyl chloride in cyclohexane

Discussion
Investigating how molecules combine with one another to form new materials is an important aspect of chemistry. In some instances, molecules bond with themselves repeatedly resulting in a remarkably long molecule called a **polymer**, which literally means "many" (poly) "units" (mer). To envision a polymer, think of molecules as paperclips. A polymer, then, is simply a long chain of paper clips linked together. When a polymer is formed from at least two different types of molecules it is called a **copolymer**. In many copolymers the different types of molecules are connected to one another in an alternate fashion—like a chain of colored paperclips that alternates from red to blue. A good example of an alternate copolymer is nylon, which is made from the molecules hexamethylenediamine and adipoyl chloride. Nylon was first prepared in the 1930's by chemists working for Du Pont, and it has since found a wide variety of applications, from nylon-stockings to guitar strings to medical sutures. Nylon was also one of the first polymers ever prepared in the laboratory. Knowledge of its synthesis led to the formulation of many other types of polymers—often called "plastics"—which are very important present-day materials.

$NH_2CH_2CH_2CH_2CH_2CH_2CH_2NH_2$

hexamethylenediamine

$+$ $(- HCl) \longrightarrow$ Nylon

$$\left(NH(CH_2)_6NH - \overset{\overset{O}{\|}}{C} - (CH_2)_4 - \overset{\overset{O}{\|}}{C} \right)_n$$

$$\underset{\text{adipoyl chloride}}{Cl \overset{\overset{O}{\|}}{C} - CH_2CH_2CH_2CH_2 - \overset{\overset{O}{\|}}{C} Cl}$$

Safety
Wear safety goggles and disposable gloves because the hexamethylenediamine (solution A) is readily absorbed through the skin. Also, this activity must be performed in a well ventilated area (preferably in a hood) as hydrogen chloride gas is formed.

A Roping Experience

Procedure

Step 1: Place about 5 mL of Solution A in a 50 mL beaker. For convenience, this quantity can be estimated using a disposable pipet. Consult your instructor.

Step 2: Slowly add the same quantity of Solution B to the 50 mL beaker. Do this in a well ventilated area because hydrochloric acid fumes (white smoke) will result. A film will form at the interface of the two solutions. Do not stir this mixture. (Why?)

Step 3: Carefully hook the film with a paper clip bent at the tip and pull the film from the beaker. It works well to hold the beaker at a slight angle and pull the nylon off the wet lip of the beaker.

Step 4: Carefully guide the nylon out of the beaker. Continue pulling until the solutions are exhausted. Lab partners should assist in managing the outcoming nylon. Try spooling the nylon around a large beaker.

Step 5: Rinse the string several times with water and lay it on a paper towel to dry. Compare the length of your nylon string to that of other groups. (Do not discard the nylon down the sink, for the drain will likely clog.)

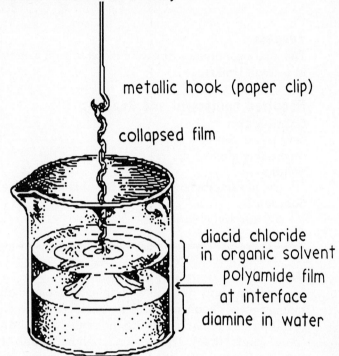

metallic hook (paper clip)

collapsed film

diacid chloride in organic solvent
polyamide film at interface
diamine in water

Summing Up

1. How long was your longest strand?_____

2. If two stands of nylon in this activity are tied together to make a longer strand, should it be counted as one's longest strand? Why or why not?

_____

3. What is a polymer? _____

_____

4. Why does Solution A contain sodium hydroxide in addition to hexamethylenediamine?

_____

5. Why is the nylon synthesized at the interface of the two liquids?

_____

6. Why is this nylon sometimes referred to as *Nylon-66*? (Hint: look at the chemical structures for hexamethylenediamine and adipoyl chloride.)

_____

A Roping Experience

CONCEPTUAL Physical Science ─────────────────── | **Activity** |

Preparation of Fragrant Esters

Smells Great!

Purpose
To prepare a variety of fragrant esters.

Required Equipment and Supplies
test tubes
test tube rack
concentrated sulfuric acid
various alcohols and carboxylic acids to be reacted together

Alcohols		Carboxylic Acids
methanol	+	salicylic acid
octanol	+	acetic acid
benzyl alcohol	+	acetic acid
isoamyl alcohol	+	acetic acid
n-propanol	+	acetic acid
methanol	+	o-aminobenzoic acid
isopentenol	+	acetic acid
methanol	+	butyric acid
isobutanol	+	propionic acid

Discussion
Many flavorings and fragrances, both natural and artificial, belong to a class of organic molecules called *esters*. These molecules contain a carbonyl group bonded to an oxygen atom that is bonded to a carbon atom (See Chapter 21, *Conceptual Physical Science*). In the laboratory, esters can be prepared by reacting alcohols with carboxylic acids (Figure 1).

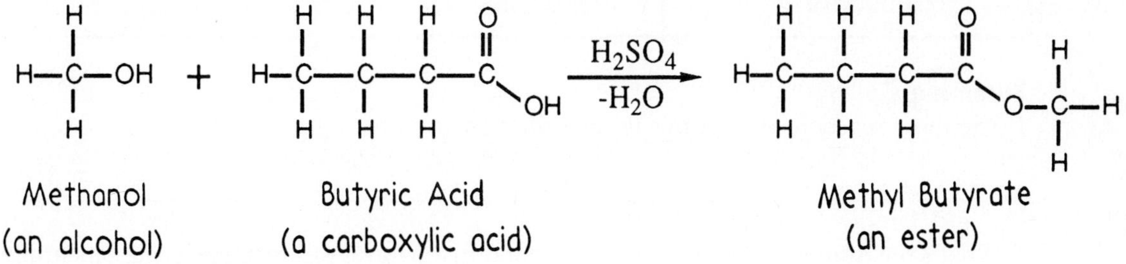

Methanol
(an alcohol)

Butyric Acid
(a carboxylic acid)

Methyl Butyrate
(an ester)

Figure 1. An example of the formation of an ester from an alcohol and a carboxylic acid.

In this activity you and your class will prepare up to nine different kinds of fragrant esters and attempt to identify their smells. To minimize the production of organic wastes each lab group will be assigned specific esters to prepare.

Safety
Your instructor will be dispensing concentrated sulfuric acid into your test tubes. Handle your samples with care. They may smell good but do not taste them — the sulfuric acid will

Smells Great!

cause severe damage to your mouth. Wear your safety goggles at all times. When you are finished with this activity empty your samples into the provided waste container. Rinse your test tubes with methanol and then clean them with soap and water.

Procedure

Step 1: Place 1 mL of each assigned alcohol along with 1 mL (or 1 gram if a solid) of its corresponding carboxylic acid in a test tube. Smell your samples before the addition of sulfuric acid. Note: it is improper and unsafe laboratory practice to stick the test tube up your nose and sniff heavily. Instead, odors can be brought to the nose by waving your hand over the lip of the test tube, which may be held several inches away from your face.

Step 2: To initiate the formation of an ester from the alcohol and the carboxylic acid, bring your samples to the instructor who will add 0.5 mL concentrated sulfuric acid. Gently tap the bottom of the test tube to promote mixing. As the ester forms a noticeable change in odor will occur. Some samples may take longer than others — allow up to 10 minutes.

Step 3: Complete Table 1 by matching the observed smell with one of the following fragrances. Check with other lab groups for samples that were not assigned to you.

Possible Fragrances:

Banana, peach, orange, grape, wintergreen, apple, rum, pear, "juicy fruit"

Table 1.

Alcohol	Carboxylic Acid	Observed Smell
methanol	salicylic acid	
octanol	acetic acid	
benzyl alcohol	acetic acid	
isoamyl alcohol	acetic acid	
n-propanol	acetic acid	
methanol	o-aminobenzoic acid	
isopentenol	acetic acid	
methanol	butyric acid	
isobutanol	propionic acid	

Summing Up

1. Are these reactions exothermic or endothermic?

_____

2. What's the difference between an ester found in a natural product, such as a pineapple, and the same ester produced in the laboratory?

_____

_____

3. Why do foods become more odorous at higher temperatures?

_____

Smells Great!

CONCEPTUAL **Physical Science**
Densities of Organic Polymers

Name that Recyclable

Purpose
To identify a variety of unknown recyclable plastics based on their densities.

Required Equipment and Supplies
recycleable plastics (PET, HDPE, LDPE, PP, PS)
solutions
    95% ethanol and water (1:1 by volume)
    95% ethanol and water (10:7 by volume)
    10% NaCl in water

Discussion
There are many different types of plastics and many of them are recycleable. Ultimately plastics need to be sorted according to their composition. For this reason plastics are coded with a number within the recycling arrow logo. The initials of the type of plastic may also appear. For example, the plastic used to make 2 liter soft-drink bottles is polyethylene terephthalate (PET). This plastic has the following recycling code:

PET

Plastics can be also be identified based upon their densities, which is useful if the recycling code is unreadable or absent. In this activity you are to identify pieces of unknown plastics based upon their densities.

Procedure
Step 1: Use the information in Table 1 to develop a separation scheme by which you will be able to identify all your unknown pieces of plastic.

Step 2: Use the solutions provided to identify each unknown piece of plastic:

Unknown No.					
Identity					

Name that Recylable

Table 1.

Recyclable Plastic	Density Compared To...			
	Water	Ethanol and Water (1:1)	Ethanol and Water (10:7)	10% NaCl in Water
1 PET — Polyethylene terephthalate	Greater	Greater	Greater	Greater
2 HDPE — High density polyethylene	Less	Greater	Greater	Less
4 LDPE — Low density polyethylene	Less	Less	Greater	Less
5 PP — Polypropylene	Less	Less	Less	Less
6 PS — Polystyrene	Greater	Greater	Greater	Less

Summing Up

1. What other physical properties might be used to identify unknown pieces of plastic?

_____

_____

2. You've just been given a thousand pounds of recyclable polystyrene and polypropylene plastic pieces all mixed together. Suggest how you might quickly separate the different types of plastic from one another.

_____

_____

Name that Recylable

CONCEPTUAL **Physical Science** ━━━━━━━━ **Experiment**
Physical and Chemical Properties

Chemical Personalities

Purpose
To observe and distinguish between the physical and chemical properties of various substances.

Required Equipment and Supplies
samples of various elements both metallic and nonmetallic
iodine crystals, sucrose crystals, and acetone
methanol
3 unknown organic solvents
copper wire
ammonium dichromate, $(NH_4)_2Cr_2O_7$
potassium chromate, K_2CrO_4
In small squeeze bottles the following aqueous reagents:
    10% sodium carbonate, Na_2CO_3
    10% sodium sulfate, Na_2SO_4
    6 M HCl
    10% calcium chloride, $CaCl_2$
test tubes, beakers, tongs, hot plates,
and Bunsen burners

Discussion
Substances can be characterized by their **physical properties**. Such properties include color, density, hardness, electrical and thermal conductivity, specific heat, and for a particular temperature, phase (solid, liquid, or gas). A **physical change** includes any change in a material substance that does not involve a change in the atomic composition of that substance. Examples of physical change occur when a substance freezes, melts, condenses, vaporizes, expands or contracts.

Substances can also be characterized by their **chemical properties**. These properties involve the tendencies of substances to undergo changes in atomic composition. For example, it is a chemical property of hydrogen to combine with oxygen to form water. It is a chemical property of iron to combine with oxygen and water to form rust. Likewise, it is a chemical property of gold to resist combining with oxygen. A change in a material involving its atomic composition is a **chemical change**. Examples include the rusting of iron, the fermenting of wine, and the burning of gasoline. During a chemical change, atoms are being pulled apart, rearranged, and put back together.

Procedure
Part A. Study of Physical Properties
Step 1: The Elements
Examine various elements and complete Table 1. Your instructor may also provide chemical compounds, which are made from elements--note their vast differences. (*Some elements are toxic. As a precaution do not open any container without the permission of your instructor.*)

Table #1

Element	Symbol	Physical Phase	Color	Metal/Nonmetal	Other Physical Properties

Step 2: Solubility

Add a few milliliters of distilled water to two small test tubes. Place a small iodine crystal in one tube and a crystal of sucrose in the other. Mix by gently tapping the bottom of the test tube with your finger. State whether the substances are *soluble, partially soluble,* or *insoluble.* Repeat the procedure using acetone as the solvent.

Iodine and water: _____   Iodine in acetone:_____

Sucrose and water: _____   Sucrose in acetone:_____

Step 3: Boiling Point

a) Construct the apparatus for the measurement of boiling point as shown below. Add about 300 mL of water to the beaker and about 2 mL of methanol, CH_3OH, to the test tube. Add a couple boiling chips to the test tube. Suspend the thermometer about 1 cm *above* the methanol. Stir the water while it is being heated and pay close attention to the thermometer readings. Note that as the methanol boils the temperature will remain constant. Record this boiling point. *(Note: Keep the methanol away from any flames.)*

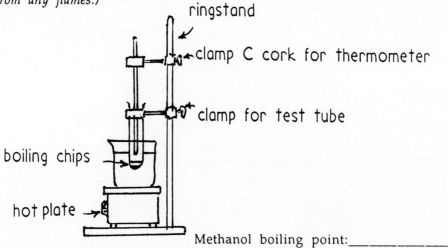

Methanol boiling point:_____

b) Obtain an unknown liquid and record its number. Repeat the above procedure for the unknown liquid and record its boiling point.

Unknown No.: _____ boiling point: _____

Part B. Study of Chemical Properties

Record your observations for Steps 4 — 8 in Table 2.

Step 4: Carefully inspect a piece of copper wire as it is heated to a glowing red hot, and then allow it to cool. Observe the changes and note whether they are physical or chemical. Specifically note where these changes are occurring.

Step 5: Place a microspatula-full of iodine crystals, I_2, in a dry beaker and cover with an evaporating dish or crucible that contains ice. In a fume hood place the beaker on a hot plate set to medium heat. State whether the change is physical or chemical. Look for any material that collects underneath the evaporating dish. This material may or may not be iodine. What evidence would you look for?

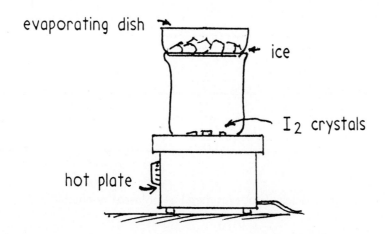

Step 6: Place less than a pea-size amount of ammonium dichromate, $(NH4)_2Cr_2O_7$, into a tall test tube, and the same amount of potassium chromate, K_2CrO_4, into a second tall test tube. In a fume hood, heat each gently over a Bunsen burner for no more than two minutes. Allow the test tubes to cool to room temperature. State whether the changes are physical or chemical. (*Note: when heating a test tube, the open end should be pointed away from you and others!*)

Step 7: Put 10 drops of sodium carbonate solution, Na_2CO_3 (aq), and sodium sulfate solution, Na_2SO_4 (aq), into separate test tubes. Add several drops of 6 M HCl and state whether the change is physical or chemical.

Step 8: Put 10 drops of sodium chloride solution, NaCl (aq), and calcium chloride solution, $CaCl_2$, into separate test tubes. Add a few drops of sodium carbonate solution, Na_2CO_3 (aq), to each of these solutions and record the changes as either physical or chemical.

Table 2

Procedure	Observations	Change
Step 4 copper + heat		
Step 5 iodine + heat		
Step 6a $(NH_4)_2Cr_2O_7$ + heat		
Step 6b K_2CrO_3 + heat		
Step 7a Na_2CO_3 + HCl		
Step 7b Na_2SO_4 + HCl		
Step 8a $NaCl$ + Na_2CO_3		
Step 8b $CaCl_2$ + Na_2CO_3		

Summing Up

1. Classify the following properties of sodium metal as physical or chemical

        a) silver metallic color   _____

        b) turns gray in air   _____

        c) melts at 98°C   _____

        d) reacts explosively with chlorine   _____

        e) dissolves in water to produce a gas   _____

        f) malleable   _____

2. Indicate whether the following observations are examples of physical or chemical changes.

        a) steam condenses to a liquid on a cool surface   _____

        b) baking soda dissolves in vinegar producing bubbles   _____

        c) mothballs gradually disappear at room temperature   _____

        d) mercury cools at -40°C to form a solid   _____

CONCEPTUAL **Physical Science**   **Experiment**

Separation of a Mixture

Sugar and Sand

Purpose
To determine the percent composition of sand in a mixture of sugar and sand.

Required Equipment and Supplies
10-gram packet of sugar and sand of unknown composition
balances measuring to the milligram
drying oven (set to 110°C)
common laboratory glassware and equipment such as:
    beakers, glass rod stirrers, spatulas, graduated cylinders, weighing papers,
    funnels, filter paper, ring stand, bunsen burner

Discussion
Components of a mixture can be separated by their differences in physical properties. In this experiment you are to isolate sand from a mixture of sugar and sand. Knowing the mass of the isolated sand and the mass of the mixture, you can calculate the percent composition of the sand that was in the mixture. Your instuctor will show you how to correctly use and care for laboratory equipment you may use, such as balances and bunsen burners. The actual procedure for finding the percent composition, however, is totally up to you. Good energy!

Safety
Remove all combustible materials, such as paper towels, from your work station when working with a bunsen burner.

Procedure
Obtain a 10 gram packet of sugar and sand from your instructor. Use available equipment to find the percent composition of sand. Be as accurate as possible. As you procede, outline everything you do and observe on a sheet of paper, which you should treat as your professional laboratory notebook.

Summing Up
1. What is the percent composition of sand in your unknown?

Unknown
Identification: [   ]   Percent
Composition: [   ]

2. Is sugar and sand an example of heterogeneous or homogeneous mixture? _____

3. If water is used in this experiment, why would distilled water be preferable to regular tap water?

_____

4. Identify several possible sources of error.

_____

Sugar and Sand

CONCEPTUAL **Physical Science** | | Activity |
Sugar Content of Soft Drinks

Sugar Soft

Purpose
To measure the sugar content of commercially-sold soft drinks using a home-built hydrometer consisting of a 9-inch plastic pipette and bolt.

Equipment and Supplies
9 inch plastic pipette
1/2 inch bolt
Standard sugar solutions (4%, 8%, 12%, 16%)
Narrow cylinder
Centimeter stick
Graph paper
Various soft drinks (Try fruit juices and diet drinks as well!)

Discussion
A hydrometer is a flotation devise used to measure the density of a liquid. Briefly, the greater the density of the liquid, the higher the hydrometer floats. In this activity, how high the hydrometer floats in four standard sugar solutions will be measured. The greater the sugar content of the solution, the greater its density, hence, the higher the hydrometer floats. A calibration curve will be graphed showing the height of the hydrometer on the y-axis and the concentration of sugar on the x-axis. How high the hydrometer floats in various commercially prepared soft drinks, which are essentially sugar solutions with small amounts of other materials, will then be measured. Using the calibration curve, the sugar content of each soft drink may be deduced.

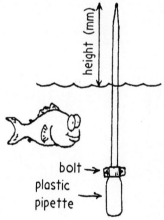

A simple hydrometer

How high the hydrometer floats out of the water is the distance between its tip and the liquid surface (use units of millimeters). Make sure that the hydrometer is not held to the sides of the container — it should float as vertically as possible. The hydrometer must be rinsed and dried before each testing. Also, carbonated beverages must be "decarbonated" since bubbles will collect on the hydrometer affecting its buoyancy. You can decarbonate a beverage by boiling it briefly in a sauce pan, then allow it to cool.

Summing Up
1. Include a table showing your tested soft drinks and their % sugar content.

2. What assumptions are made in this activity, and how might you test their validity?

_____

_____

CONCEPTUAL **Physical Science**   **Experiment**

Molecular Models

Molecules by Acme

Purpose
For this experiment you will build models of various molecules. First, you are given a chemical formula, then, based upon some rules for how atoms bond, you are to piece together a molecular model.

Required Equipment and Supplies
Molecular modeling kit. Students should work in groups of 2 or 3.

Discussion
The 3-dimensional shapes of molecules can be envisioned with the use of molecular models. There are certain things you must know, however, before you can assemble an accurate model of a molecule. First, you need to know what types of atoms make up the molecule and also their relative numbers. This information is given by the molecule's chemical formula. The chemical formula for water, H_2O, for example, tells us that each and every water molecule is made of two hydrogen atoms and one oxygen atom. The second thing you need to know is how the atoms of the molecule fit together. For a water molecule there are several possibilities. One hydrogen atom might be bonded to both the second hydrogen atom and the oxygen atom (Figure 1a), or all three atoms might be bonded in the shape of a three membered ring (Figure 1b). We find through chemistry, however, that there are specific ways in which different types of atoms bond. Hydrogen, for example, bonds once, while oxygen bonds twice when it can. Knowing this we can build a more reasonable model of a water molecule where both hydrogen atoms are bonded once to a central oxygen atom (Figure 1c).

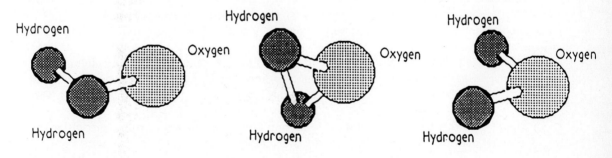

Hydrogen   Oxygen   Hydrogen   Oxygen   Hydrogen   Oxygen

Hydrogen   Hydrogen   Hydrogen

a   b   c

Figure 1. Which version of H_2O is correct?

The shape of a molecule is largely responsible for the physical and chemical properties of the molecule. If water were linear like carbon dioxide, for example, its boiling point would be close to that of carbon dioxide, –78°C. Doesn't this mean the Earth's oceans would be gaseous?

Molecules by Acme

The Molecular Model

Molecular modeling kits represent different types of atoms with different colored pieces. Hydrogen atoms, for example, are typically represented by white pieces and oxygen atoms by red pieces. Also, the number of times an atom prefers to bond is indicated by the number of times the piece is able to attach to other pieces. Hydrogen atoms, for example, bond only once, so hydrogen pieces have only one site for attachment. Similarly, oxygen pieces have two sites for attachment. To build a molecule, the pieces representing atoms are connected by sticks (or springs). You will know when you have built a correct structure for a molecule when each atom is bonded the appropriate number of times (Table 1). In some instances, you may find it necessary to form multiple bonds between atoms. Consult the instructions to your modeling kit or your instructor to see how this is done. This is a hands-on, play-as-you-investigate laboratory. Enjoy!

Table 1.

Bonds* Type of Atom	Atomic Symbol	Typical color of sphere	Number of holes
Hydrogen	[H]	white	1
Carbon	[C]	black	4
Nitrogen	[N]	blue	3
Oxygen	[O]	red	2
Chlorine	[Cl]	green	1

* The number of times an atom tends to bond is related to its position in the periodic table. Consider, for example, the relative positions of carbon, nitrogen, oxygen and chlorine. That these atoms are in adjacent columns and prefer 4, 3, 2 and 1 bonds, respectively, is not a coincidence. The periodic table is more than a table of facts. With further study you will find that the periodic table is highly organized — and a lot like a roadmap, for much information about an element can be told merely from its position.

Procedure

Step 1: Get a set of models. You may wish to work with a partner. The sets may not contain equal numbers of pieces, so you may occasionally need to borrow from other groups.

Step 2: Determine which colors should be used to represent the following elements: hydrogen, carbon, nitrogen, oxygen, chlorine, and iron. The number of bonds they are able to form should be as listed in Table 1. Enter the actual colors of the pieces you use in Table 2.

Step 3: Build models of each of the following molecules. Avoid forming triangular rings made of 3 atoms. They are strained and less stable. Check your structure with your instructor — in several cases there is more than one possibility. Complete Table 3 by drawing an accurate representation of your structure (follow the example of the water molecule.) Answer the end-of-activity questions using these models.

1. Hydrogen gas.......... H_2
2. Oxygen gas............. O_2
3. Nitrogen gas........... N_2
4. Water.................... H_2O
5. Hydrogen Peroxide... H_2O_2
6. Ammonia............... NH_3
7. Methane................ CH_4
8. Dichloromethane....... CH_2Cl_2

9. Chloroethanol........ C_2H_5ClO
10. Carbon Dioxide...... CO_2
11. Acetylene............. C_2H_2
12. Ethanol................ C_2H_6O
13. Acetic Acid........... $C_2H_4O_2$
14. Benzene.............. C_6H_6
15. Iron(III) Oxide....... Fe_2O_3

Molecules by Acme

Table 2.

Element:	Hydrogen [H]	Carbon [C]	Nitrogen [N]	Oxygen [O]	Chlorine [Cl]	Iron (III)* [Fe]
Color:						

* For iron, choose a piece that is able to bond in 6 directions at angles of 90 degrees.
The iron will form only 3 bonds such that 3 potential bonding sites remain unconnected.

Table 3.

Hydrogen H_2	Oxygen O_2	Nitrogen N_2	Water H_2O	Hydrogen Peroxide H_2O_2
			H O H	

Ammonia NH_3	Methane CH_4	Dichloromethane CH_2Cl_2	Chloroethanol C_2H_5ClO	Carbon Dioxide CO_2

Acetylene C_2H_2	Ethanol C_2H_6O	Acetic Acid $C_2H_4O_2$	Benzene C_6H_6	Iron (III) Oxide Fe_2O_3

Molecules by Acme

Summing Up

1. a. The Principle of Definite Proportions states that atoms combine to form molecules in definite ratios. In a water molecule, for example, there are two hydrogens for every one oxygen. If this ratio were different, say two hydrogens to two oxygens, would the shape of the molecule also be different?

_____

   b. Would you still have a water molecule?

_____

2. a. How are the structures for methane and dichloromethane similar?

_____

_____

   b. How are they different?

_____

_____

3. How many *different* structures are possible for the formula C_2H_5ClO?

_____

4. Of the 15 models you made, which are linear?

_____

5. Which molecules have multiple bonds between atoms?

_____

6. The flat structure of one of these molecules is partly responsible for it being a cancer causing substance. Because it is flat, this molecule can literally slice into the DNA double helix, thereby messing up the cell's ability to regulate itself. Which molecule is this?

_____

Molecules by Acme

CONCEPTUAL **Physical Science** ━━━━━━━━━━━━━━━ **Experiment**

Qualitative Analysis

Mystery Powders

Purpose
To identify 10 common household chemicals by using qualitative analysis, a method for identifying substances through a process of elimination.

Required Equipment and Supplies

Household chemicals and formulae
cornstarch, $(C_6H_{10}O_5)_n$
white chalk, $CaCO_3$
plaster of Paris, $2CaSO_4 \cdot H_2O$
washing soda, Na_2CO_3
lye, $NaOH$
epsom salt, $MgSO_4 \cdot 7H_2O$
baking soda, $NaHCO_3$
boric acid, H_3BO_3
table sugar, $C_{12}H_{22}O_{11}$
table salt, $NaCl$

Test reagents
tincture of iodine
phenolphthalein
white vinegar
rubbing (isopropyl) alcohol (70%)
sodium hydroxide (0.3 M)

Equipment
test tubes
eye dropper
beaker
spatula

Discussion
You are given ten vials and each contains a white powder, which is a common household chemical. Your task is to identify these unknowns based upon their different physical and chemical properties. For this experiment you should develop a qualitative analysis scheme, such as shown in Figure 1, to show how the chemicals can be systematically identified.

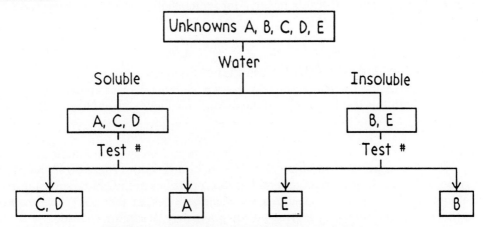

Figure 1. A sample qualitative analysis scheme

Safety
Do not taste any of the unknowns. Some may taste sweet but others may burn a hole through your tongue! Treat all the unknowns with caution and respect. Avoid spillage. Use small quantities — no more than is required for each test. Never place excess unknown back in the reagent bottle, for this can contaminate the stock material. Consult your instructor for proper disposal.

Mystery Powders

Procedure

Using the tests outlined below prepare a qualitative analysis scheme that should permit you to unequivocally determine the identities of the unknowns. This scheme should be prepared before conducting this experiment. It is recommended that you start with Test #1 (Solubility in Water). You may then choose your own order of testing. Some orders are more efficient than others. Try to develop a scheme that minimizes the number of test tubes you must use. Use Table 1 as a guide to the physical and chemical properties of the household chemicals to be identified.

Tests

#1 Solubility in Water: place a pea-sized portion of the unknown solid in a test tube and add about 5 mL of water. To mix the contents, hold the top of the test tube securely between your thumb and index finger and gently slap the bottom of the test tube with the index finger of your opposite hand. Be careful so that none of the solution spills out. Consult Table 1 for behaviors of each of the household chemicals.

#2 Tincture of Iodine: This test may be used for any unknown that is insoluble in water. Add a few drops of tincture of iodine to the unknown as it sits undissolved in water. A deep blue color forms as the iodine complexes with cornstarch. A brownish color will appear for all other unknowns insoluble in water.

#3 Phenolphthalein: This test may be used for any unknown that is soluble in water. Add a drop of phenolphthalein to the dissolved unknown and a bright pink color will result if the solution is alkaline (pH > 7). This test is positive for lye, NaOH, and washing soda, Na_2CO_3.

#4 White Vinegar: This test is applicable to all unknowns. Add a few drops of vinegar to the unknown, in either a dissolved or undissolved state. The formation of bubbles is a sign of the carbonate ion [CO_3^{2-}], which decomposes to gaseous carbon dioxide upon treatment with an acid (vinegar). This test is positive for white chalk, $CaCO_3$, washing soda, Na_2CO_3, and baking soda, $NaHCO_3$.

#5 Sodium Hydroxide (0.3 M): This is a specific test for magnesium sulfate, $MgSO_4$, which is found in epsom salts. If the unknown dissolves in water and forms an insoluble precipitate when treated with this test reagent, then the unknown contains magnesium sulfate.

#6 Hot Water: All the water soluble unknowns become markedly more soluble in warmer water with the exception of sodium chloride, NaCl (table salt). This test, therefore, is specific to sodium chloride. Place several pea-sized portions of the unknown solid in a test tube along with 5 mL of water. Heat in a hot water bath held at about 60°C. No marked improvement in solubility suggests that the unknown may be sodium chloride.

#7 Rubbing Alcohol: This test may be used for any unknown that is soluble in water. Place a pea-sized portion of the unknown solid in a test tube and add about 7 mL of rubbing alcohol (isopropyl, 70%). Only three substances should dissolve readily: epsom salt, $MgSO_4$, lye, NaOH, and boric acid, H_3BO_3. Four substances will not dissolve readily: baking soda, $NaHCO_3$, washing soda, Na_2CO_3, sugar, $C_{12}H_{22}O_{11}$, and salt, NaCl.

Mystery Powders

Table 1. Properties of 10 Household Chemicals

Household Chemical	#1 Solubility	#2 Iodine	#3 pH >7	#4 Vinegar	#5 Hydroxide	#6 Hot water	#7 Alcohol
			Reactions to Tests				
cornstarch	insoluble	positive	--	negative	--	--	--
white chalk	insoluble	negative	--	positive	--	--	--
plaster of Paris	insoluble	negative	--	negative	--	--	--
washing soda	soluble	--	positive	positive	negative	greater sol.	insoluble
lye	soluble	--	positive	negative	negative	greater sol.	soluble
epsom salt	soluble	--	negative	negative	positive	greater sol.	soluble
baking soda	soluble	--	negative	positive	negative	greater sol.	insoluble
boric acid	soluble	--	negative	negative	negative	greater sol.	soluble
table sugar	soluble	--	negative	negative	negative	greater sol.	insoluble
table salt	soluble	--	negative	negative	negative	no effect	insoluble

Summing Up

1. In the following table specify your unknowns.

Household Chemical:	corn-starch	white chalk	plaster of Paris	washing soda	lye	epsom salt	baking soda	boric acid	table sugar	table salt
Unknown Identification:										

2. Which of the tests used in this experiment measure physical properties and which measure chemical properties?

_____

3. Sugar is a chemical, but it is also a food. Is this contradictory? Briefly explain.

_____

_____

4. How many individual tests did you have to perform to identify all of the unknowns? (Review your qualitative analysis scheme or simply count your number of dirty test tubes.)

_____

5. Is it more efficient to start with tests that split the group of chemicals in half, such as Test #7, or to start with tests that are specific for single compounds, such as Test #5?

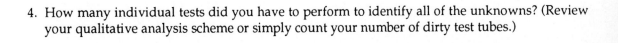

Mystery Powders

CONCEPTUAL **Physical Science**

Red Cabbage Juice pH Indicator

Sensing pH

Purpose

To isolate and use the pH sensitive pigment of red cabbage.

Supplies

Red Cabbage
Cooking pot
Water

Discussion

The pH of a solution can be approximated by means of a pH indicator — any chemical whose color changes with pH. Many pH indicators are found in plants, a good example being red cabbage.

Procedure

Shred about a quarter of a head of red cabbage. Boil the shredded cabbage in 2 cups of water for about 5 minutes. Strain the cabbage while collecting the broth, which contains the pH indicator. Red cabbage indicator is red at low pHs (pH = 1-4), a light purple at neutral pHs (pH = 7), green at moderately alkaline pHs (pH = 8-11), and yellow at very alkaline pHs (pH = 13). Add small amounts of cabbage broth to various solutions, such as white vinegar, rain water, ammonia, or bleach, to estimate their pHs.

Sensing pH

CONCEPTUAL Physical Science — **Experiment**

Acid/Base Titration

Upset Stomach

Purpose
To measure and compare the acid neutralizing strengths of three antacids.

Required Equipment and Supplies
50.00 mL burets (2)
buret stand
50 mL beakers (2)
250 mL Erlenmeyer flasks (3)
balances
mortar and pestle
Antacids: Rolaids™; Tums™; Alka-Seltzer™
0.50 M HCl (aq)
0.50 M NaOH (aq)
phenolpthalein pH indicator
universal pH paper

Discussion
Overindulgence of food or drink can lead to acid indigestion, a discomforting ailment that results from the excess excretion of hydrochloric acid, HCl, by the stomach wall lining. An immediate remedy is an over-the-counter antacid, which consists of a base that effectively neutralizes the acid. In this experiment, antacid will be added to a simulated upset stomach — an Erlenmeyer flask filled with 30 mL of acid. Not all of the acid, however, will be neutralized. Your task is to find out how much remains. If a lot remains, then the antacid did not neutralize very well. On the other hand, if only a little acid remains, then the antacid was more effective. How much acid remains will be found by completing the neutralization with another base, sodium hydroxide, NaOH.

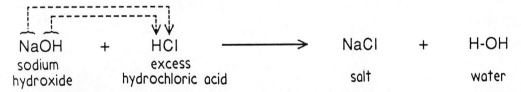

Recall that during a chemical reaction atoms are neither created nor destroyed. Instead, they just change partners. We see this in the neutralization reaction between NaOH and HCl, where sodium and chlorine atoms come together to form sodium chloride, NaCl (table salt), and hydrogen and oxygen atoms come together to form water, H_2O. The product of this reaction and many other acid/base reactions is plain old neutral salt water!

Safety
Safety goggles should be worn at all times when working with the hydrochloric acid, HCl, and sodium hydroxide, NaOH, solutions. If either gets on your skin, wash thoroughly with running water. The HCl solution may sting after a few minutes, and the NaOH solution will immediately feel very slippery.

Procedure

Step 1: Using a buret, deliver 30 mL of a 0.50 M solution of hydrochloric acid, HCl, to a 250 mL Erlenmeyer flask. Be sure to read the liquid level from the bottom of the meniscus. Record initial and final buret readings in the Data Sheet and calculate the actual volume of HCl delivered. (The buret should be read to 2 decimal places, for example, 29.93 mL — the hundredth's digit must be estimated by reading the buret carefully). This flask with acid in it represents a stomach with acid indigestion. Add 3 to 5 drops (no more) of phenolphthalein indicator to the acid. The solution should remain colorless.

Step 2: Crush and grind an antacid tablet with a mortar and pestle and carefully transfer the resulting fine powder to a weigh-dish. Measure the mass of the powder on a balance. (Note: the Alka-Seltzer need not be crushed.) Carefully transfer the antacid to the "stomach flask", and swirl the stomach carefully for a few minutes. Measure the mass of the empty weigh-dish and calculate the mass of antacid that was added to the stomach-flask. Record this information in the Data Sheet.

Step 3: Using universal pH paper, note the pH of the "stomach fluid" after treatment with the antacid. Use a glass rod or spatula to transfer a small drop from the flask to the paper.

You can find the number of milliliters of HCl that were neutralized by the antacid by carefully neutralizing any remaining HCl with the 0.50 M solution of sodium hydroxide, NaOH. When just enough NaOH is added to neutralize all the remaining HCl, the solution will turn a light pink color due to the phenolphthalein indicator added in Step 1. If you add too much NaOH, the solution will turn a dark pink. If you add too little NaOH, the pink color will disappear within 30 seconds. You should add enough NaOH such that the light pink color persists for at least a minute. Since the concentration of the NaOH is the same as the original stomach acid (0.05 M), the volume of NaOH used to neutralize the HCl is equal to the volume of the HCl that was not neutralized by the antacid.

Step 4: Record the initial volume of a buret containing at least 30 mL of a 0.50 M solution of NaOH. Carefully deliver this solution to the "relieved stomach fluid" in increments. Initially you will see a pink color form and then fade away as you swirl the flask. As you get closer to the point of complete neutralization (called the *end-point*) the pink color will persist for longer periods of time. Be sure to add the NaOH more slowly as you approach the end-point so as not to "overshoot". Record the final volume of the buret after you have reached the end-point and calculate the total volume of NaOH required to neutralize all remaining hydrochloric acid.

Step 5: Repeat this procedure for the remaining antacid samples to be tested.

Summing Up

1. Compare the antacids based upon the number of milliliters of acid they neutralized per gram and list them in order of neutralizing strength.

Most Strong _____ > _____ > _____ Least Strong

2. Predict the color of the litmus paper in Step 3 if only 10 mL of acid were initially delivered at the beginning of each procedure. (Predict either "red" for acidic, "tan" for neutral, or "blue" for basic).

<u>circle one</u>

Rolaids' (1 tablet):	red	tan	blue
Rolaids' (2 tablets):	red	tan	blue
Tums' (1 tablet):	red	tan	blue
Tums' (2 tablets):	red	tan	blue
Alka-Seltzer' (1 tablet):	red	tan	blue
Alka-Seltzer' (2 tablets):	red	tan	blue

Consider making photocopies of the Titration Data Sheet on following page 130 so you'll have one sheet for each antacid.

Titration Data Sheet
(one per antacid)

Antacid: _____   Active Ingredient: _____

Step 1: Delivery of 0.50 M HCl to 250 mL flask ("stomach").

Final volume of acid: _____ mL

Initial volume of acid: _____ mL
_____

Total volume of acid: _____ mL

Read the buret to the hundredths place!

Number of drops of phenolphthalein indicator added: _____

Step 2: Mass of crushed antacid that was delivered to the "stomach": _____ g
(Alka-Seltzer' need not be crushed)

> Use this area for any miscellaneous calculations

Step 3: pH of "relieved stomach fluid" (using universal pH paper): _____

Step 4: Delivery of 0.50 M NaOH to HCl treated with antacid.

...where the solution turns a permanent light color pink

Final volume of base: _____ mL

Initial volume of base: _____ mL
_____

Total volume of base: _____ mL

What was the number of milliliters of 0.50 HCl that the antacid neutralized? _____ mL

How many milliliters of acid were neutralized for every <u>gram</u> of antacid

_____ mL/g

This last question is useful because it allows us to compare antacids on a mass-to-mass basis.

CONCEPTUAL Physical Science ━━━━━━━━━━━━━━━━━ **Experiment**

Rocks and Minerals

Crystal Growth

Purpose
To observe the growth of crystals from a melt and from a solution.

Required Equipment and Supplies
thymol
sodium chloride
sodium nitrate
potassium aluminum sulfate (alum)
copper acetate
glass slide or clear glass plate
forceps
petri dish
hot plate
microscope

Discussion
A mineral is a naturally formed inorganic solid composed of an ordered array of atoms. The atoms in different minerals are arranged in their own characteristic ways. This systematic arrangement of atoms is known as a mineral's crystalline structure, which exists throughout the entire mineral specimen. If crystallization occurs under ideal conditions, it will be expressed in perfect crystal faces.

I: Growth of Crystals from a Melt
Most minerals originate from a magma melt. Crystalline minerals form when molten rock cools. The temperature at which minerals crystallize is very high (well above the boiling point of water) making direct observation difficult. In this experiment we will use thymol, an organic chemical that crystallizes near room temperature. Although thymol is different from a magma, the crystallization principles exhibited are similar.

> **Safety** Thymol is not poisonous but may cause skin and eye irritation. Be safe: Use forceps to handle thymol.

Procedure
Part A. Slow Cooling Without "Seed" Crystals
Set hot plate to low heat. Place a petri dish containing a small amount of crystalline thymol on the hot plate until all the crystals are melted. Allow the melt to heat for one to two minutes and then set it aside to cool slowly. Do not disturb during cooling. This melt will be examined at the end of the lab session.

Part B. Slow Cooling With "Seed" Crystals

Repeat above procedure (A) but transfer the petri dish to the stage of a microscope as soon as the thymol is melted. Add several (four or five) "seed" crystals to the melt and observe it under the microscope. As the melt cools, you will see the crystals begin to grow.

Describe the manner of crystal growth. Think about the rate of growth, the direction of growth, the crystal faces, and the effect of limited space on crystal shape. What role do the "seed" crystals play in initiating crystal growth?

_____

_____

3. Make a sketch of the final crystalline solid. The crystals produced in this experiment are similar to the type of crystals found in common igneous rocks. Both are interlocking crystals. Compare the thymol crystals to those in a rock specimen of granite. Sketch the texture of the granite. How do the thymol crystals compare with those in granite?

_____

_____

Part C. Rapid Cooling

Repeat above procedure (A) but transfer the petri dish to the top of an ice cube and let it cool for 30 seconds. The melt will cool very rapidly. Quickly move the petri dish to the microscope and observe the nature of the crystal growth. Time permitting, repeat this procedure until you are sure of your observations and can state your general conclusions.

_____

_____

Crystal Growth

Part D. Examination of Crystals from Procedure Part A

Examine the thymol crystals from the first part of this exercise. Observe the crystal size. What effect does the rate of cooling have on the crystal size?

_____

_____

II: Growth of Crystals from Solution

Many minerals are precipitated from aqueous solutions by evaporation. Crystal growth of such minerals can be observed in the laboratory by evaporating prepared concentrated solutions.

Procedure

Obtain concentrated solutions of sodium chloride, sodium nitrate, potassium aluminum sulfate (alum), and copper acetate from your instructor.

Place a drop of each solution on a microscope slide. Label each slide accordingly. As the water evaporates from the slide, crystals of each compound will appear. With time and continued evaporation, the crystals will grow larger.

Summing up

1. How does crystallization from an aqueous solution compare to crystallization from a melt?

_____

_____

2. Do the crystals precipitated from a solution have unique crystal forms?

_____

_____

3. On the basis of this experiment, do you think that crystal forms provide a good reference for mineral identification?

_____

_____

4. On the basis of this experiment, what is the relationship between cooling and crystal size?

_____

_____

Crystal Growth

CONCEPTUAL **Physical Science**

Rocks and Minerals

What's that Mineral?

Purpose
To observe the physical properties of various mineral samples and identify them by a systematic procedure.

Required Equipment and Supplies
mineral collection
hardness set — piece of glass, steel knife, copper penny,
    streak plate (non-glazed porcelain plate)
dilute hydrochloric acid (HCl)

Discussion
A mineral is a naturally formed inorganic solid with a characteristic chemical composition and crystalline structure. The different combinations of its elements and arrangement of atoms determine the physical properties of a mineral: its shape, the way it reflects light, its color, hardness, and its mass.

I. The physical properties dependent on a mineral's chemical composition include luster, streak, color, specific gravity, and reaction with acid.

The **luster** of a mineral is the appearance of its surface when it reflects light. Luster is independent of color; minerals of the same color may have different lusters and minerals of the same luster may have different colors. Mineral luster is classified as either metallic or nonmetallic.

Test for Luster
Metallic minerals are usually
        a. gold, silver or black
        b. shiny, polished
        c. opaque
        d. always have a streak
Nonmetallic minerals are usually
        a. not gold or silver
        b. shiny to dull
        c. transparent, translucent, or opaque
        d. rarely have a streak

The **streak** of a mineral is the color of its powdered form. We can see a mineral's streak by rubbing it across a nonglazed porcelain plate. Minerals with a metallic luster generally leave a dark streak that may be different from the color of the mineral. For example, the mineral hematite is normally reddish-brown to black, but always streaks red. Magnetite is normally iron-black, but streaks black. Limonite is normally yellowish-brown to dark brown, but always

What's that Mineral?

streaks yellowish-brown. Minerals with a nonmetallic luster either leave a light streak or no streak at all.

```
Test for Streak
1. Scrape the mineral across a nonglazed porcelain plate.
2. Blow away excess powder.
3. The color of the powder is the streak.
```

Although **color** is an obvious feature of a mineral, it is not a reliable means of identification. When used with luster and streak, color can sometimes aid in the identification of metallic minerals. Nonmetallic minerals may occur in a variety of colors or be colorless. Therefore, color is not used for the identification of nonmetallic minerals. In this exercise we will only differentiate between light-colored and dark-colored minerals.

Specific gravity, (s.g.), is the ratio of the mass (or weight) of a substance to the mass (or weight) of an equal volume of water. Metallic minerals tend to have a higher specific gravity than nonmetallic minerals. For example, the metallic mineral gold (Au) has a specific gravity of 19.3 whereas quartz (SiO_2), a nonmetallic mineral, has a specific gravity of 2.65.

The **reaction to acid** is an important chemical property often used to identify carbonate minerals. Carbonate minerals effervesce (fizz) in dilute hydrochloric acid (HCl). Some carbonate minerals react more readily with HCl than others. For instance, calcite ($CaCO_3$) strongly effervesces when exposed to HCl but dolomite ($CaMg(CO_3)_2$) doesn't react unless it is scratched and powdered.

II. The physical properties dependent on a mineral's crystalline structure include hardness, cleavage and fracture, crystal form, striations, and magnetism.

The resistance of a mineral to being scratched (or its ability to scratch) is a measure of the mineral's **hardness**. The varying degrees of hardness are represented by Mohs scale of hardness. For this activity we are concerned with the hardness of some common objects.

Cleavage and fracture are useful guides for identifying minerals. **Cleavage** is the tendency of a mineral to break along planes of weakness. Planes of weakness depend on crystal structure and symmetry. Some minerals have distinct cleavage. Mica, for example, has perfect cleavage in one direction, and breaks apart to form thin, flat sheets. Calcite, has perfect cleavage in three directions, and breaks to produce rhombohedral faces that intersect at 75-degrees. A break other than along cleavage planes is a **fracture**.

Mohs Scale of Hardness		
Mineral	Scale number	Common objects
Diamond	10	
Corundum	9	
Topaz	8	
Quartz	7	Steel file
Orthoclase	6	
		Window glass or Pocket knife
Apatite	5	
Fluorite	4	
Calcite	3	Copper penny
Gypsum	2	Fingernail
Talc	1	

What's that Mineral?

Test for Hardness
1. Place a glass plate on a hard flat surface.
2. Scrape the mineral across the glass plate.
3. If the glass is scratched, the mineral is harder than glass.
4. If the glass is not scratched, the mineral is softer than glass.
5. If the mineral is softer than glass, test to see if it is harder than a copper penny. Scrape the mineral across the penny.
6. If the mineral is softer than the penny, test to see if it is harder than your fingernail. Try to scratch the mineral with your fingernail.

Every mineral has its own characteristic **crystal form**. Some minerals have such a unique crystal form that identification is relatively easy. The mineral pyrite, for example, commonly forms as intergrown cubes, while quartz commonly forms as six-sided prisms that terminate in a point. Most minerals, however, do not exhibit their characteristic crystal form. Perfect crystals are rare in nature because minerals typically grow in cramped, confined spaces.

Cleavage Patterns

Number of cleavage directions	Shape	Number of flat surfaces	
1	Flat sheets	2	
2 at 90°	Rectangular cross section	4	
2 not at 90°	Parallelogram cross section	4	
3 at 90°	Cube	6	
3 not at 90°	Rhombohedron	6	
Fracture	Irregular shape	0	

Some minerals have grooves on their cleavage planes. These grooves, called **striations**, can be used to differentiate between feldspar minerals. Plagioclase feldspars have straight parallel striations on one cleavage plane. Orthoclase feldspars have lines that resemble striations but are actually color variations within the mineral. These grooves are not straight, and they are not parallel to each another. Instead, these "striations" make a criss-cross pattern.

Some minerals exhibit magnetism. To test for magnetism simply expose the mineral to a small magnet or a compass.

So we see that physical properties that depend on a mineral's crystalline structure include hardness, crystal form, cleavage and fracture, striations, and magnetism.

What's that Mineral?

Minerals with Metallic Luster

Streak color	Properties	Mineral
Black to gray	silver gray 3 directions of cleavage at 90° cubic crystals hardness = 2.5 specific gravity = 7.6	Galena
Black to gray	black to dark gray magnetic hardness = 6 specific gravity = 5.	Magnetite
Black to gray	gray to black marks paper feels greasy hardness = 1-2 specific gravity = 2.2	Graphite
Black to greenish black	golden yellow may tarnish purple hardness = 4 specific gravity = 4.2	Chalcopyrite
	brass yellow may tarnish green cubic crystals striations hardness = 6 specific gravity = 5.2	Pyrite (fool's gold)
Reddish brown to black	silver to gray may tarnish reddish brown hardness = 5-6 specific gravity = 5	Hematite
Yellowish brown to reddish brown	black to silver gray to golden brown may tarnish yellowish brown hardness = 5.5 specific gravity = 4	Limonite

Some are interesting...
some are more interesting!

What's that Mineral?

Minerals with Nonmetallic Luster — Dark Colored

Hardness	Cleavage	Properties	Mineral
Harder than glass	Present	black to blue gray 2 directions of cleavage, not quite at 90° striations on one cleavage plane hardness = 6 specific gravity = 2.7	Plagioclase
		dark green to black 2 directions of cleavage at nearly 90° no striations hardness = 6 specific gravity = 3.5	Pyroxene
		dark green to black 2 directions of cleavage intersecting at 60° and 120° no striations hexagonal crystals hardness = 5.5 specific gravity = 3.6	Amphibole (hornblende)
Softer than glass		brown to dark green to black 1 direction of cleavage producing thin sheets translucent hardness = 3 specific gravity = 3	Biotite
		green to dark green 1 direction of cleavage producing thin curved sheets greasy luster hardness = 2.5 specific gravity = 2.9	Chlorite

Earth science is down to earth!

What's that Mineral?

Minerals with Nonmetallic Luster — Dark Colored

Hardness	Cleavage	Properties	Mineral
Harder than glass	Absent	olive green to black no streak glassy luster conchoidal fracture hardness = 7 specific gravity = 2.6	Olivine
		light to dark gray no streak glassy, transparent, translucent conchoidal fracture hexagonal crystals hardness = 7 specific gravity = 2.6	Quartz
		deep red to brown no streak translucent conchoidal fracture isometric crystals hardness = 7 specific gravity = 4	Garnet
Variable		red reddish brown streak opaque, earthy uneven fracture hardness = 1.5 - 5.5 specific gravity = 5.2	Hematite

When I dream of geology
I have rocks in my head!

What's that Mineral?

Minerals with Nonmetallic Luster — Light Colored

Hardness	Cleavage	Properties	Mineral
Harder than glass	Present	flesh pink, white, green, brown 2 cleavage directions at nearly 90° no striations color lines on cleavage planes hardness = 6 <u>specific gravity = 2.6</u>	Orthoclase feldspar
Harder than glass	Present	white to blue gray 2 directions of cleavage not quite at 90° striations on one cleavage plane hardness = 6 <u>specific gravity = 2.7</u>	Plagioclase feldspar
Softer than glass	Present	colorless to white 3 directions of cleavage at 90° soluble in water hardness = 2.5 <u>specific gravity = 2.2</u>	Halite
Softer than glass	Present	colorless to white 1 direction of cleavage hardness = 2 (easily scratched with fingernail) <u>specific gravity = 2.3</u>	Gypsum
Softer than glass	Present	colorless to white or yellow 3 directions of cleavage not at 90° (rhomb shaped) translucent to transparent strong reaction to acid hardness = 3 specific gravity = 2.7	Calcite

Hmmm...gold maybe?

What's that Mineral?

Minerals with Nonmetallic Luster — Light Colored

Hardness	Cleavage	Properties	Mineral
Softer than glass	Present	white, gray, pink 3 directions of cleavage not at 90° (rhomb shaped) opaque reacts to acid when powdered hardness = 3.5 specific gravity = 2.9	Dolomite
		yellow, blue, green, violet 4 directions of cleavage transparent to translucent cubic crystals hardness = 4 specific gravity = 3.2	Fluorite
		colorless to pale green 1 direction of cleavage producing thin elastic sheets hardness = 2.5 specific gravity = 2.7	Muscovite
		white to greenish 1 direction of cleavage pearly luster hardness = 1 specific gravity = 2.8	Talc
Harder than glass	Absent	white, gray, pink, violet glassy luster conchoidal fracture hardness = 7 specific gravity = 2.6	Quartz
		olive green to yellow green glassy luster conchoidal fracture hardness = 7 specific gravity = 3.5	Olivine

What's that Mineral?

Procedure
A. Mineral Identification

Examine the various unknown minerals and note their characteristics on the following worksheet. Identify the different minerals by comparing your list to the mineral identification tables.

Separate metallic and nonmetallic minerals
- A. If mineral is metallic determine:
  1. streak
  2. color
  3. hardness
  4. any other distinguishing properties
  5. name the mineral
- B. If the mineral is nonmetallic determine:
  1. color (separate light from dark minerals)
  2. hardness
  3. cleavage
  4. any other distinguishing properties
  5. name the mineral

Mineral Identification Worksheet

Luster	Streak	Color	Cleavage	Fracture	Hardness	Specific Gravity	Other Characteristics	Mineral Name

What's that Mineral?

Summing Up

1. What distinguishing characteristic is used in identifying the following minerals?

   a) halite _____

   b) pyrite _____

   c) quartz _____

   d) biotite _____

   e) fluorite _____

   f) garnet _____

2. Indicate whether the following physical properties result from a mineral's crystalline structure or a mineral's chemical composition.

   a) crystal form _____

   b) color _____

   c) cleavage _____

   d) specific gravity _____

3. If a mineral does not exhibit a streak, is it metallic? Explain.

   _____

   _____

4. Which property is more reliable in mineral identification, color or streak? Why?

   _____

   _____

5. What physical properties distinguish biotite from muscovite?

   _____

   _____

6. What physical properties distinguish plagioclase feldspars from orthoclase feldspars?

   _____

   _____

What's that Mineral?

CONCEPTUAL **Physical Science** | Activity

Rocks and Minerals

Rock Hunt

Purpose
To practice finding geology everywhere.

Required Equipment and Supplies
Desire

Discussion
Activity 1
Go on a rock hunt. Now that you are more familiar with rocks and minerals you can start your own collection. Gather rocks from the beach, an old river channel, a stream bed, a roadcut, or even your back yard. Collect at least six different looking rocks and classify them into the three major rock groups. What tell-tale features help you classify the rocks? Can you identify your rocks by these features?

Activity 2
Rocks are not only found on the beach and in the mountains, but almost everywhere. In fact, if you live in a city you are probably surrounded by more rock materials than you realize. Take a field trip down any city street and you will notice that most buildings are constructed from stone material. Many stone buildings are polished providing an easy view of a rocks mineral composition, texture, and hence its method of formation. Is the rock composed of visible crystals? Are the crystals interlocking? Are the crystals flattened or deformed? What is the grain size? Are there fossils? Support your classification with your observations.

Activity 3
Observe the buildings in your locality. Look for the older buildings, most of which were built from local material. The different rock types used in the construction of some of these buildings can tell you much about local history. Is there a marble or granite quarry in your area? Try to trace the building material to its origin.

Rock Hunt

CONCEPTUAL **Physical Science** | **Activity** |

● *Rocks and Minerals*

What's that Rock?

Purpose
To identify rocks from the three different types: igneous, metamorphic, and sedimentary.

Required Equipment and Supplies
collection of assorted igneous, sedimentary, and metamorphic rocks
dilute hydrochloric acid (HCl)

Discussion
The three major classes of rocks — igneous, sedimentary, and metamorphic — have their own distinct physical characteristics. By learning to identify the representative characteristics of each rock type we may gain a better understanding of the history recorded in the earth's crust.

Procedure
You will be given three sets of rocks — igneous, sedimentary, and metamorphic. Your task is to identify the rocks in each set. The first set to identify are igneous rocks. Refer to Table 1, *Classification of Igneous Rocks* to aid in your identification. The second group of rocks are the sedimentary rocks. For this part of the exercise refer to Table 2, *Classification of Sedimentary Rocks.* To make identification and classification easier, the table has been divided into two parts — Part A for clastic sedimentary rocks, and Part B for nonclastic sedimentary rocks. The final group of rocks you will identify are the metamorphic rocks. Refer to Table 3, *Classification of Metamorphic Rocks* to help distinguish the different metamorphic rocks.

Part A: Identification of Igneous Rocks
Molten magma welling up from within the earth produces igneous rock of two types — *intrusive* and *extrusive*. Intrusive rocks are formed from magma that solidified below the earth's surface and extrusive rocks are formed from magma erupted at the earth's surface.

Step 1: The first step in identifying igneous rocks is to observe their texture. Texture is related to the cooling rate of magma in a rock's formation. Magma that solidifies below the earth's surface cools slowly, forming large, visible interlocking crystals that can be identified with the unaided eye. This coarse-grained texture is described as **phaneritic**. If the texture is exceptionally coarse-grained (visible minerals larger than your thumb) the rock is described as having a **pegmatitic** texture. Some intrusive rocks contain two distinctly different crystal sizes in which some minerals are conspicuously larger than the other minerals. This texture, with big crystals in a finer groundmass, is called **porphyritic**.

In contrast, magma that reaches the surface tends to cool rapidly forming very fine-grained rocks. This fine-grained texture is described as **aphanitic**. When magma is cooled so quickly that there is not time for the atoms to form crystals, the texture is described as **glassy**. Many aphanitic rocks contain cavities left by gases escaping from the rapidly cooling magma. These gas bubble cavities are called vesicles, and the rocks that contain them are said to have a **vesicular** texture. If the gaseous magma cools very quickly the texture that develops is described as **frothy** (foam-like glass). In a volcanic eruption, the forceful escape of gases causes rock fragments to be torn from the sides of the volcanic vent. These rock fragments combine with volcanic ash and cinders to produce a **pyroclastic**, or fragmental texture.

What's that Rock?

Classification of Igneous Rocks

Composition / Texture	Light-Colored	Intermediate-Color	Dark-Colored	Very Dark Color
	10-20% quartz K-feldspar>plagioclase ≈10% ferromagnesian minerals	No quartz plagioclase>K-feldspar 25-40% ferromagnesian minerals	No quartz plagioclase≈50% 50% ferromagnesian minerals	100% ferromagnesian minerals
Pegmatitic (very coarse-grained)	Pegmatite			
Porphyritic (mixed crystal sizes)	Porphyry			
Phaneritic (coarse-grained)	Granite / Granodiorite	Diorite	Gabbro	
Aphanitic (fine-grained)	Rhyolite	Andesite	Basalt	Peridotite / Dunite
Glassy	Obsidian			
Porous (glassy, frothy)	Pumice		Scoria	
Pyroclastic (fragmental)	Volcanic tuff (fragments < 4mm) Volcanic breccia (fragments > 4mm)			

Table 1

Step 2: Observe the color, and hence the chemical composition of an igneous rock sample. Igneous rocks are either rich in silica or poor in silica. Silica-rich (quartz-rich) rocks tend to be light in color whereas silica-poor rocks tend to be dark in color. Igneous rocks can therefore be divided into a light-colored group — quartz, feldspars, and muscovite; and a dark-colored group — the ferromagnesian minerals — biotite, pyroxene, hornblende, and olivine. (*Ferromagnesian* refers to minerals that contain iron [ferro] and magnesium [magnesian]).

Part B: Identification of Sedimentary Rocks

Rock material that has been weathered, subjected to erosion, and eventually consolidated into new rock is sedimentary rock. Sedimentary rocks are classified into two types — clastic and nonclastic. **Clastic** sedimentary rocks are formed from the compaction and cementation of fragmented rocks. **Nonclastic** sedimentary rocks are formed from the precipitation of minerals in a solution. This chemical process can occur directly, as a result of inorganic processes, or indirectly, as a result of biochemical reaction.

Step 1: Observe the texture of the rock. If the rock is composed of visible particle grains the rock is probably clastic. Clastic sedimentary rocks are classified according to particle size. Large sized particles range from boulders to cobbles to pebbles. If the large-sized particles are rounded, the rock is a **conglomerate**. If they are angular and uneven, the rock is a **breccia**. Medium sized particles produce the various types of sandstone. **Quartz sandstone** is composed wholly of well rounded and well sorted quartz grains. **Arkose** is composed of quartz with about 25 percent feldspar grains. **Graywacke** is composed of both quartz and feldspar grains with fragments of broken rock in a clay type matrix. Silts and clays comprise the fine sized particles. Silt sized particles produce **siltstone** and clay sized particles produce **shale**.

What's that Rock?

Step 2: If the rock does not fit into the clastic classification, the rock is probably nonclastic. Nonclastic sedimentary rocks are classified by their chemical composition. Inorganic processes produce the **evaporites** — minerals formed by the evaporation of saline water. The evaporites include **rock salt** and **gypsum**. Rock salt can be identified by its salty taste. Biochemical reactions produce **carbonates** by way of calcareous-secreting organisms, and **chert**, by silica-secreting organisms. **Limestone, chalk,** and **dolostone** are carbonate rocks. Recall that carbonate minerals react (fizz) to HCL. Chert can be identified by its waxy luster and conchoidal fracture. One rock, **coal** (a biogenic sedimentary rock) is formed from the accumulation and compaction of vegetation matter.

Classification of Clastic Sedimentary Rocks

Particle size		Rock Name	Characteristics
Coarse (> 2mm)		Conglomerate	Rounded grain fragments
		Breccia	Angular grain fragments
Medium (1/16 - 2mm)	Sandstone	Quartz	Quartz grains (often well rounded, well sorted)
		Arkose	Quartz and feldspar grains (often reddish color)
		Graywacke	Quartz grains, small rock fragments, and clay minerals (often grayish color)
Fine (1/256 - 1/16 mm)		Siltstone	Silt sized particles, surface is slightly gritty
(<1/256 mm)		Shale	Clay sized particles, surface has smooth feel, no grit

Classification of Non-clastic Sedimentary Rocks

Composition		Rock Name	Characteristics
Halite (NaCl)	Evaporites	Rock Salt	Salty taste
Gypsum (CaSO$_4$*2H$_2$O)		Gypsum	Inorganic precipitate
Organic matter		Coal	Compacted, carbonized plant remains
Calcium Carbonate (CaCO$_3$)		Limestone	Fine-grained, strong reaction to HCl
		Chalk	Microfossils; fine-grained
		Fossiliferous limestone	Macrofossils and fossil fragments
Dolomite (CaMg(CO$_3$)$_2$)		Dolostone	Reaction to HCl when in powdered form
Quartz (SiO$_2$)		Chert	Hard, dense, waxy luster, may show conchoidal fracture

Table 2

Part C: Identification of Metamorphic Rocks

Rock material that has been changed in form by high temperature or pressure is metamorphic rock. Metamorphic rocks are most easily classified and identified by their texture; and when a particular mineral is very obvious, by their mineralogy. Metamorphic rocks can be divided into two groups: **foliated** and **nonfoliated**.

Foliated metamorphic rocks have a directional texture and a layered appearance. The most common foliated metamorphic rocks are slate, phyllite, schist, and gneiss

Slate is very fine-grained and is composed of minute mica flakes. The most noteworthy

What's that Rock?

characteristic of slate is its excellent rock cleavage — the property by which a rock breaks into plate-like fragments along flat planes. **Phyllite** is composed of very fine crystals of either muscovite or chlorite that are larger than those in slate, but not large enough to be clearly identified. Phyllite is distinguished from slate by its glossy sheen. **Schists** have a very distinctive texture with a parallel arrangement of the sheet structured minerals (mica, chlorite, and/or biotite). The minerals in a schist are often large enough to be easily identified with the naked eye. Because of this, schists are often named according to the major minerals in the rock (biotite schist, staurolite-garnet schist, etc.). **Gneiss** contains mostly granular, rather than platy minerals. The most common minerals found in gneiss are quartz and feldspar. The foliation in this case is due to the segregation of light and dark minerals rather than alignment of platy minerals. Gneiss has a composition very similar to granite and is often derived from granite.

Nonfoliated rocks are mono-mineralic and thus lack any directional texture. Their texture can be described as coarsely crystalline. Common nonfoliated metamorphic rocks are **marble** and quartzite.

Classification of Metamorphic Rocks

Foliated Metamorphic Rock

Crystal Size		Rock Name	Characteristics
Very fine, crystals not visible		Slate	Excellent rock cleavage
Fine grains, crystals not visible		Phyllite	Well developed foliation; glossy sheen
Coarse Texture	S	Muscovite schist	
	c	Chlorite schist	
Crystals visible with unaided eye	h	Biotite schist	Mineral content reflects
Micaceous minerals	i	Garnet schist	increasing metamorphism
Often contains large crystals	s	Staurolite schist	from top to bottom
	t	Kyanite schist	
		Sillimanite schist	
Coarse		Gneiss	Banding of light and dark minerals

Nonfoliated Metamorphic Rock

Precursor Rock	Rock Name	Characteristics
Quartz sandstone	Quartzite	Interlocking quartz grains
Limestone	Marble	Interlocking calcite grains

Table 3

Summing Up

1. Did you find evidence that igneous rocks can exhibit both fine and coarse textures? Describe the textures.

_____

_____

_____

_____

What's that Rock?

2. What are the influencing factors for large crystal formation? What types of rock exhibit enlarged crystals?

_____

_____

_____

3. What distinguishing characteristics are exhibited in sedimentary rocks?

_____

_____

_____

4. How can we distinguish igneous rocks from metamorphic rocks?

_____

_____

_____

Going Further

For this exercise you will first determine if the rock is igneous, sedimentary, or metamorphic. Then you will use the classification tables for the different rock types to identify the rocks. Examine your rock specimen closely. Look at its texture. Can you see individual mineral grains? If so, the texture is coarse to medium coarse. If the mineral grains are too small to be identified, the texture is fine. How are the grains arranged? Are the grains interlocking? Interlocking grains are formed by crystallization, so the rock is probably either igneous or metamorphic. Are the interlocking grains aligned? Do they show foliation? Foliation indicates metamorphism. Are the grains separated by irregular spaces filled with cementing material? Are fossils present? Does the rock react with acid? If so, the rock is sedimentary. If the rock is crystalline and shows no reaction to acid, check the hardness of the rock. Metamorphic and igneous rocks are harder than sedimentary rocks.

What's that Rock?

CONCEPTUAL **Physical Science**

Specific Gravity of Rocks

Specific Gravity

Purpose
To measure the specific gravity of various rock samples.

Required Equipment and Supplies
rock and metal samples
electronic balance
beaker of water

Discussion
The density of an object is its mass/volume. The specific gravity of an object is the ratio of its density to the density of water — or the ratio of the weight of an object to the weight of an equal volume of water.

$$\text{Specific Gravity} = \frac{\text{weight of object}}{\text{weight of equal volume of water}}$$

A rock that weighs 4 times as much as an equal volume of water, for example, has a specific gravity of 4. Specific gravities of various substances are shown in the table below.

An object submerged in water displaces its own volume of water. Recall from Chapter 5 that the buoyant force that acts on an object submerged in a fluid is equal to the weight of the fluid displaced — Archimedes Principle. The weight of this displaced water is the buoyant force. So we can say

$$\text{Specific Gravity} = \frac{\text{weight of object in air}}{\text{buoyant force}}$$

Force is measured in newtons, but because weight and mass are proportional, and specific gravity is a ratio, we can express buoyant force in grams.

Procedure
Place a sample rock on the measuring pan of an electronic balance and record its mass.

Mass of rock in grams = _____

Remove the rock and place a beaker of water on the measuring pan. Tare this mass — that is, set the reading of the balance to zero. Submerge the sample rock in the water by suspending it by a thin string. The balance will read the buoyant force (the weight of water displaced by the rock).

Buoyant force in grams = _____

Specific gravity of rock = _____

Do the same for other samples.

Summing Up
1. Why are there no units for specific gravity? _____

2. Would your calculated specific gravities differ if you expressed your weight in ounces instead of grams? Why or why not?

_____

Specific Gravity of Various Substances	
Ice	0.92
Water	1.00
Borax	1.7
Quartz	2.65
Concrete	2.7
Aluminum	2.70
Talc	2.8
Mica	3.0
Olivine	3.6
Chromite	4.6
Pyrite	5.0
Hematite	5.26
Steel	7.8
Iron	7.86
Brass	8.44
Copper	8.93
Silver	10.5
Gold	19.3

Specific Gravity

CONCEPTUAL **Physical Science** | Activity

Surface Features — Mapping

Top This

Purpose
To interpret maps, particularly topographic maps, which show landforms and approximate elevations above sea level. Using points of known elevation you will learn to draw contour lines and using a topographic map you will learn to construct a topographic profile.

Required Equipment and Supplies
topographic quadrangle map (provided by your instructor)
ruler
pencil and eraser

Discussion
Maps provide a representation of the earth's surface and are a very useful tool. A map is a scaled down, idealized representation of the real world. Everything on a map must be proportionally smaller than what it really is. Roads, waterways, mountains, ground area, and distance are all proportionally reduced in scale. A map's **scale** is defined as the relationship between distance on a map and distance on the ground. There are three ways to describe scale on a map. A *graphical scale* is a drawing, a line marked off with distance units. A *verbal scale* uses words to describe distance units — "one inch to one mile". The third way is a *representative fraction* (rf) that gives the proportion as a fraction or ratio, such as 1:24,000. This means that one unit of measure on the map — 1 inch or 1 centimeter — is equal to 24,000 units of the same measure on the ground. If the scale is 1:10,000 then 1 inch on the map is equal to 10,000 inches on the ground. The first number refers to the map distance and is always 1. The second number refers to the ground distance and will change depending on scale.

1:24,000-scale   1:100,000-scale   1:125,000-scale

Fig. 1 U.S. Geological Survey Maps of the same area at different scales. The scale of a map tells us how much area is being represented and the level of detail in that area. When comparing maps, we see that the smaller the scale the larger the area represented, and the larger the scale the smaller the area represented.

Top This

Large-Scale Map *Example: 1:10,000*	Small-Scale Map *Example: 1:1,000,000*
Shows a small area with a lot of detail.	Shows a large area with very little detail.
A large scale is good for urban, street or hiking maps that require detail.	A small scale is good for world or regional maps that cover a large area.

Scale Conversion

It is difficult to envision 10,000 inches, let alone 1,000,000 inches on the ground. In order to make these units more meaningful we can convert them into miles or kilometers.

Question   What is 1 inch on a 1:25,000 scale map in reality (on the ground)?

Answer   Using conversion   1 foot = 12 inches
                      5280 feet = 1 mile
                       1 mile = 1.609 kilometers

$$25{,}000 \text{ in } \times \frac{1 \text{ ft}}{12 \text{ in}} \times \frac{1 \text{ mile}}{5280 \text{ ft}} = .394 \text{ miles}$$

$$.394 \text{ miles} \times \frac{1.609 \text{ km}}{1 \text{ mile}} = .634 \text{ kilometers}$$

Question   What is 1 mile on a 1:25,000 scale map in inches?

Answer $1 \text{ mi } \times \dfrac{5280 \text{ ft}}{1 \text{ mi}} \times \dfrac{12 \text{ in}}{1 \text{ ft}} \times \dfrac{1 \text{ in on map}}{25000 \text{ in}} = 2.53 \text{ inches on map}$

Relief Portrayal and Topographic Maps

A **topographic map** is a two-dimensional representation of a three-dimensional land surface. Topographic maps show **relief** — the extent to which an area is flat or hilly. To show land surface form and vertical relief, topographic maps use contours. A **contour** line connects all points on the map having the same elevation above sea level. Each contour line acts to separate the areas above that elevation from the areas below it (Figures 2 and 3).

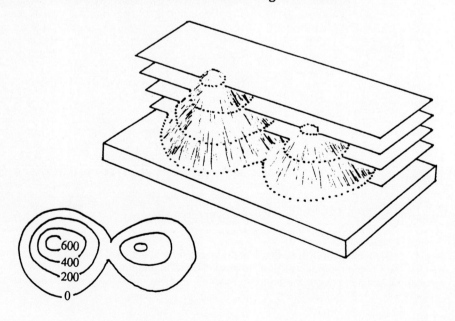

Fig. 2

Top This

162

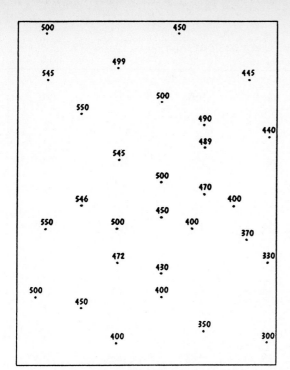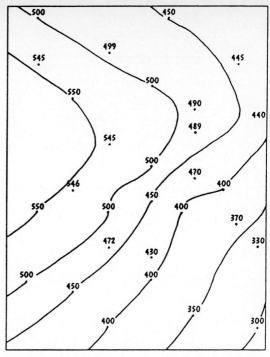

Fig. 3 Known points of elevation are shown at the left. Contour lines from known points of elevation as shown at the right. The contour interval = 50 feet.

The vertical difference in elevation between contour lines is called the **contour interval**. The contour interval is usually stated at the lower margin of a topographic map. If it is not stated, look for a numbered contour line, an *index contour line* that is thicker and darker. Every fifth line is usually an index line. The choice of contour interval depends on the scale of the map. All contour lines are multiples of the contour interval. For a contour interval of 10 feet the sequence of contours would be 0, 10, 20, 30, 40, etc. For a contour interval of 50 feet the sequence of contours would be 0, 50, 100, 150, 200, etc. If a point lies on a numbered contour line, the elevation is simply read from the line. If the contour line is unnumbered, the elevation can be found by counting the number of lines (contour intervals) above or below the index contour. Add or subtract this number from the index contour. If a point lies between contour lines, the elevation is approximated from the two contour lines it lies between. Elevations of specific points, such as mountain peaks, road crossings, or survey stations can sometimes be read directly from the map. Most of these points are marked with a small "x" and a number for the elevation.

Rules for Contour Lines
1. All points on a contour line must be of equal elevation.
2. Contour lines always close to form a closed path, like an irregular circle. If the contour line extends beyond the mapped area the entire circle will not be seen.
3. Contour lines never cross one another.
4. Contour lines must never divide or split.
5. Closely spaced contour lines represent a steep slope, widely spaced contour lines represent a gentle slope.
6. A concentric series of closed contours represents a hill.
7. A concentric series of closed contours with hachure marks directed inward represent a closed depression.
8. Contour lines form a V pattern when crossing streams. The apex of the V always points upstream (uphill).

Top This

Summing Up Exercises

1. Draw in all necessary contour lines below. Use a contour interval of 10 meters.

135	122	110	88	100	105	90
140	125	113	94	120	121	100
143	132	120	108	127	135	140
150	135	128	120	140	144	155
156	138	134	138	147	156	165
160	146	141	153	155	165	180

2. Draw in all necessary contour lines below. Use a contour interval of 25 meters.

165	124	75	63	123	187	198
132	100	99	67	71	114	148
116	148	169	102	60	63	67
115	176	197	142	100	69	54
101	147	206	187	149	113	70
93	123	200	203	148	87	159
82	98	154	205	185	193	175
60	101	166	169	156	153	199
56	95	164	132	141	160	173

Top This

3. Mark each contour line with its corresponding elevation. Use a contour interval of 40 feet.

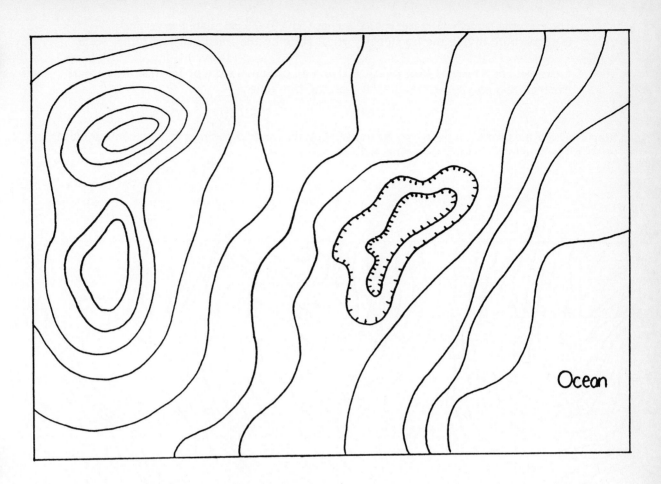

Topographic Profile

A topographic map provides an overhead view of the landscape. A profile of the landscape across any area of the map can be constructed by drawing a straight line across the desired area. This profile is like a slice through the landscape allowing us a side view. Figure 4 shows a method for constructing a topographic profile.

Procedure

Step 1: Mark a straight line across the map to indicate the line of profile; lable it A-A'. Placement of the line can be anywhere, depending on what part of the landscape you want to view.

Step 2: Lay the edge of a sheet of paper along the line A-A'. Mark the position of each contour on the paper. Note the elevation for each mark. Other features such as mountain crests or the location of streams should also be marked.

Step 3: Prepare a graph with the line A-A' as the horizontal axis and elevation as the vertical axis. Each line will represent a contour line. Each line on the graph must be equally spaced. Label the lines so that the highest and lowest elevations along the line of profile fit into the graph. The vertical axis can be made to the same scale as the map or, for a more impressive profile, the scale can be exaggerated. Note: the size of exaggeration must be stated for correct interpretation of the profile.

Top This

For vertical exaggeration, first decide on a vertical scale.

For example: 1 inch = 100 feet = 1200 inches (1:1200). To calculate the vertical exaggeration, divide the fractional vertical scale by the fractional horizontal scale.

Example: For a vertical scale of 1:10,000 and a horizontal scale of 1:50,000 the vertical exaggeration will be 5 times greater than the true relief.

Step 4: Draw the profile on the graph by transferring the marked position of each contour elevation on its respective line. Connect the points with a smooth line.

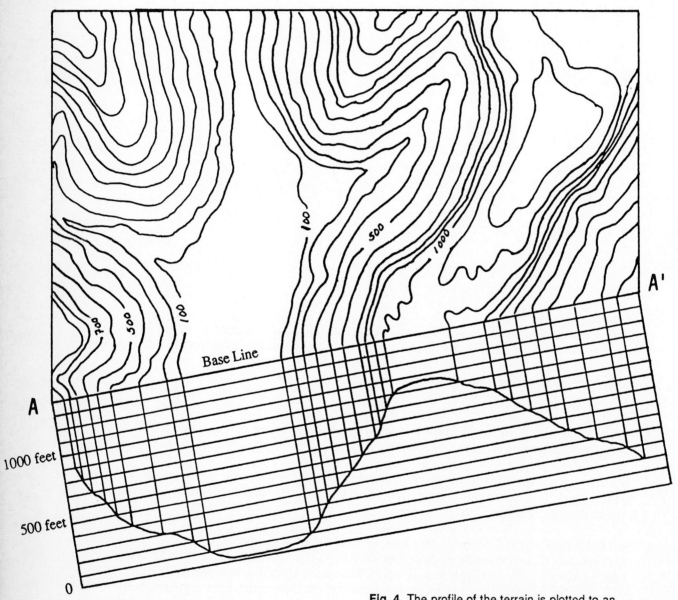

Fig. 4 The profile of the terrain is plotted to an appropriate scale from the chosen baseline.

Top This

166

Exercise 4. Construct a topographic profile for line A-A' below. Calculate the vertical exaggeration.

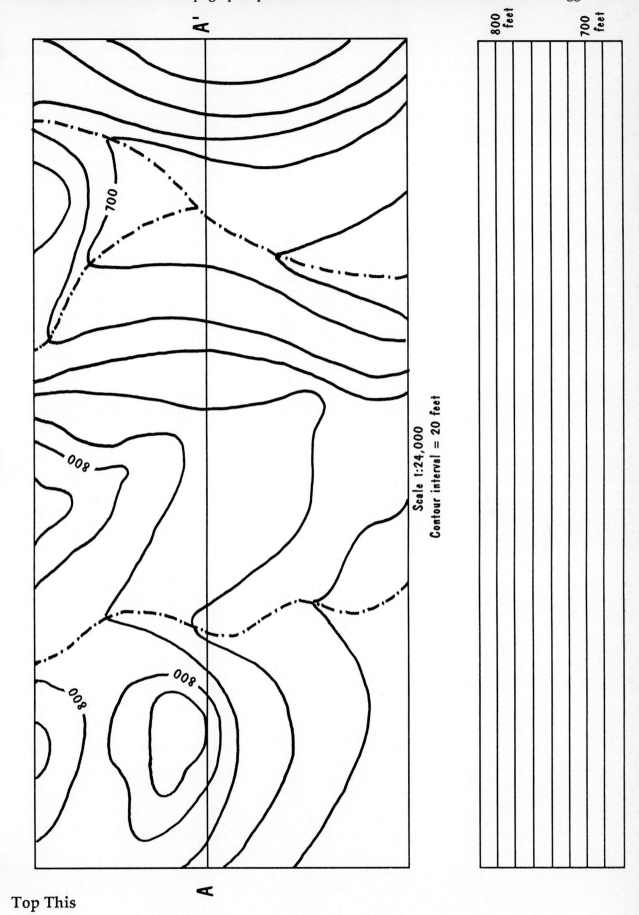

Scale 1:24,000
Contour interval = 20 feet

Top This

CONCEPTUAL Physical Science ━━━━━━━━━━━━━━━━━━━━ | Activity |

Surface Features — Mapping

Water Below Our Feet

Purpose

To construct contour maps of the watertable and determine flow paths. Data for map constuction can be water table levels obtained from wells and other techniques. You will construct contour lines from data. Your contour lines represent the water table in much the same way as contour lines represent land surface.

Required Equipment/Supplies

ruler
pencil and eraser
protractor

━━━━━━━━━━━━━━━━━━━━━━━━━━━━━━━━━━━━━━━

Discussion

Below the surface of continents is an extensive and accessible reservoir of fresh water. This reservoir of subsurface water is divided into two classes — soil moisture and groundwater. Groundwater is water that has percolated into the subsurface and saturated the open pore spaces in the rock or soil. The upper boundary of this saturated zone is called the **water table**. Water in the unsaturated zone above the water table is called *soil moisture*. The depth of the water table varies with precipitation and climate.

The water table tends to be a subdued version of the surface topography — sort of an underground surface. Watertable contour maps, similar to surface contour maps, show the direction and speed of groundwater flow — information extremely useful for water supply management. For instance, simulated water table elevations generated by computer models can be compared with actual water table elevations in order to calibrate and verify the model. A groundwater model will tell you the best location for a well, and the impact that pumping from a new well will have on current water levels. Of particular interest today is the use of groundwater modeling to monitor contaminant transport in the subsurface.

Groundwater flow is influenced by the force of gravity. Groundwater flows "downhill" underground, but the path it takes is dependent on *hydraulic head* and not topography. Hydraulic head is higher where the water table is high, such as beneath a hill, and lower where the water table is low, such as beneath a stream valley. So, responding to the force of gravity, water moves from high areas of the water table to low areas of the water table. As groundwater flows, it takes the path of least resistance — the shortest route. For any two lines of equal elevation, the shortest distance between them will be perpendicular to the lines. So, if we think of the path of groundwater flow as a *flow line*, the flow line will always be perpendicular to the contour lines.

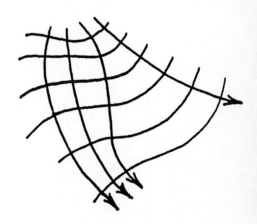

Problem 1: Construction of Water Table Contour Lines

The construction of contour lines for a water table is very similar to the construction of surface topographic contour lines. The same procedures and rules for surface contours apply to water table contour maps. The only difference is that you are drawing contour lines of the water table below the ground surface.

Draw contour lines for the elevation of the water table. Use a contour interval = 10 feet.

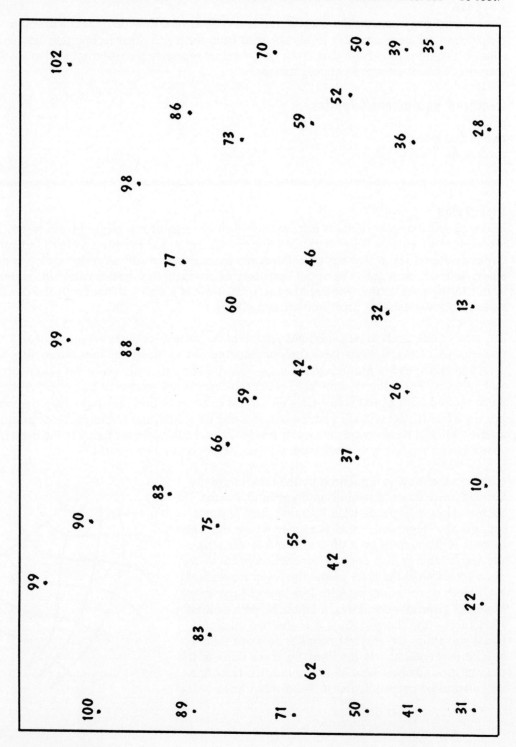

Water Below Our Feet

Problem 2: Construction of Flow Lines

Water flows from areas of high hydraulic head to areas of lower hydraulic head. In mapping groundwater flow the key concept is that the line of flow is perpendicular to the contour lines of the water table. You are provided with the cross section view as a guide to your thinking.

Draw flow lines on the water table contour maps, with arrows to show flow direction. Also draw arrows on the surface of the water table, at cross sections to indicate the direction of flow.

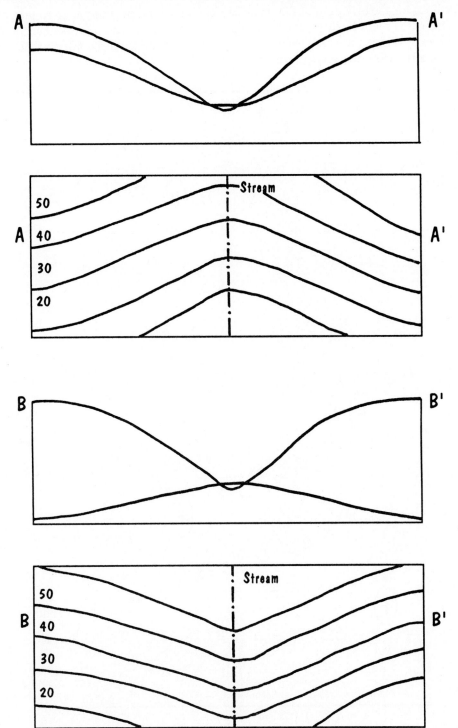

Water Below Our Feet

A knowledge of groundwater flow is important for solving problems of groundwater contamination. The most common groundwater contamination comes from sewage — drainage from septic tanks, inadequate or leaking sewer pipes, and farm waste areas. Contamination may also result from landfills and waste dumps where toxics and hazardous wastes leach down into the subsurface.

Problem 3: Contaminant Flow Model

Draw contour lines for the elevation of the water table. Use a contour interval of 10 feet. Water wells and a sewer treatment facility have been plotted on the map. The treatment facility is in disrepair and has developed a leak. Determine which well (if any) will be affected by the flow of contamination from the sewage treatment facility.

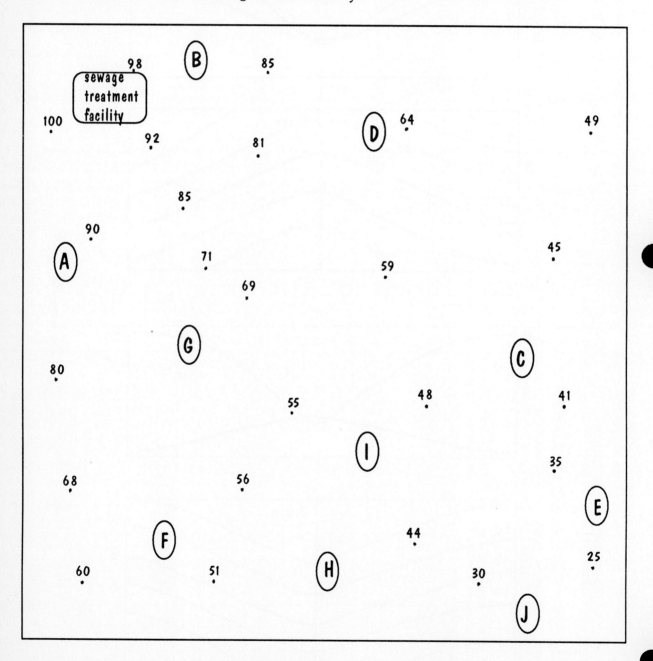

Water Below Our Feet

Summing Up

1. In Problem 2, what is the difference between stream A and stream B? In what type of climatic area would we likely find stream A? In what type of climatic area would we likely find stream B? Explain.

_____

_____

_____

2. In Problem 3 we are assuming equal rates of pumping for each well. If excessive pumping occurred at well H, would there be a change in the flow of contamination? Which well, if any, would likely become contaminated because of this pumping?

_____

_____

_____

3. In Problem 3 we are assuming that all the soil in the subsurface is homogeneous. If an impermeable clay lens is found at point 55, would the flow of sewage contamination be affected? How about point 35?

_____

_____

_____

Water Below Our Feet

Darcy's Law

Purpose
To verify Darcy's Law and determine the hydraulic conductivity of soil samples.

Required Equipment and Supplies
pen and pencil, and a calculator
graph paper

Discussion
In the mid nineteenth century the French engineer Henry Darcy discovered that the rate of groundwater flow is directly proportional to both *hydraulic conductivity* and the *hydraulic gradient*—the difference in hydraulic head of two points divided by the distance between them. Darcy's formula is

$$q = K \times i$$

where q is the flow rate, K is the hydraulic conductivity, and i is the hydraulic gradient. Darcy packed a cylinder with sand and measured how fast water flowed through it.

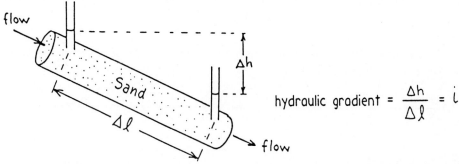

Procedure
We can recreate the spirit of Darcy's experiment by analyzing some hypothetical laboratory data. The first set of data shows the relationship between flow rate through a sample of sand and hydraulic gradient. After verifying Darcy's Law we can use his formula to determine the hydraulic conductivities of different sands (or other geologic material).

On a separate sheet of graph paper, plot the following data. Let the vertical axis be Flow Rate (cm/s) and the horizontal axis Hydraulic Gradient (cm/cm).

Flow Rate (cm/s)	Hydraulic Gradient (cm/cm)
6.0×10^{-4}	0.2
1.2×10^{-3}	0.4
1.8×10^{-3}	0.6
2.4×10^{-3}	0.8
3.0×10^{-3}	1.0

Darcy's Law

Summing Up

1. If Darcy's Law is valid for the above data, the plot should be a straight line on the graph of flow rate versus hydraulic gradient. Does Darcy's Law hold?

_____

2. What is the hydraulic conductivity of this sand?

_____

3. The following data represent five experiments to determine the hydraulic conductivities of five different sands. Each experiment was performed using a hydraulic gradient of 0.6. Calculate the hydraulic conductivity of each sand and complete the table.

Sand	Flow Rate (cm/s)	Hydraulic Conductivity (cm/s)
1	6.0×10^{-4}	
2	1.2×10^{-3}	
3	1.8×10^{-3}	
4	2.4×10^{-3}	
5	3.0×10^{-3}	

4. Darcy's actual experiment related the volumetric flow rate Q to the cross sectional area A of the cylinder in addition to the hydraulic conductivity and hydraulic gradient:

$$Q = A \times K \times i$$

For a cylinder of 10 cm² cross section, find Q for a sand sample with a hydraulic gradient of 0.8 and a flow rate of 5.0×10^{-3} cm/s.

_____

Going Further

Q tells us the volume of water that flows out of the cylinder in a given amount of time. Its units, for example, are cm³/s. q is more like a velocity, with units cm/s. Neither of these quantities tell us the average speed of a particular water molecule or a contaminant moving through the cylinder. To know this value, we need to know the *effective* cross sectional area of the cylinder. This depends on the *porosity* of the geologic material. (Like squeezing a garden hose, where the speed of the water increases but the volume of water coming out of the hose in a given amount of time stays the same — conservation of mass.)

Two different sediment samples may have the same hydraulic conductivity but different porosities. This may occur if one sample is poorly sorted and the other is well sorted. Question: Would the average velocity of a water molecule be higher or lower if porosity is increased while holding hydraulic conductivity constant? Defend your answer.

_____

_____

CONCEPTUAL **Physical Science** | Activity

Earth's Internal Properties

Over and Under

Purpose
To construct and interpret maps of geologic cross sections in the subsurface.

Required Equipment and Supplies
protractor
compass
colored pencils and eraser

Discussion
Geologic cross sections show the three dimensional structure of the subsurface. They provide a "pie slice" of the subsurface and are thus extremely useful for mineral, ore, and oil exploration. We can construct geologic cross sections by using surface information — outcrops of folded and faulted rock sequences, and igneous rock intrusions.

The study of rock deformation is called *structural geology*. When a rock is subjected to compressive stress it begins to buckle and fold. If the stress overcomes the strength of the rock, the rock exhibits strain and breaks or faults. Structural geology interprets the different types of stress and strain.

We measure the orientation of deformed rock layers using strike and dip (Figure 1). Strike is the trend or direction of a horizontal line in an inclined plane. The direction of strike is expressed relative to north. For example, "north X degrees west" or "north X degrees east". Dip is the vertical angle between the horizontal plane and an inclined plane. Dip is always

Fig. 1 Strike and dip. On the rock outcrop, the strike is the line formed from the intersection of a horizontal plane and the tilted rock strata. The dip is the angle between the horizontal and tilted stata (plane). The direction of dip is simply the geograpical (N,S,E, or W) direction in which a marble would roll down the tilted plane. Dip is always measured perpendicular to strike. In the example shown, the outcrop is striking *northwest* and dipping *45° southwest.*

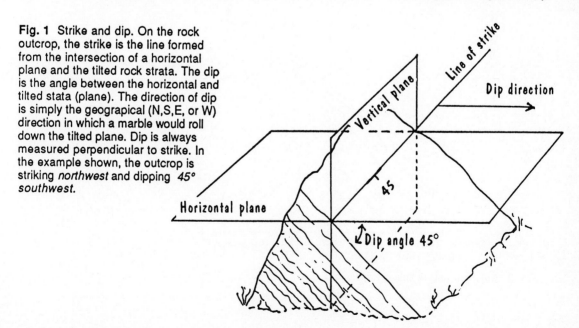

measured perpendicular to strike. We can think of the direction of dip as the direction a marble would roll down a plane. Hence, we express dip as (1) the direction a marble would roll, and (2) the angle between the inclined and horizontal planes. In Figure 1 the dip direction and angle

Over and Under   177

measured perpendicular to strike. We can think of the direction of dip as the direction a marble would roll down a plane. Hence, we express dip as (1) the direction a marble would roll, and (2) the angle between the inclined and horizontal planes. In Figure 1 the dip direction and angle are expressed as "45° west". Geologic map symbols and symbols for strike and dip are shown in Figure 2.

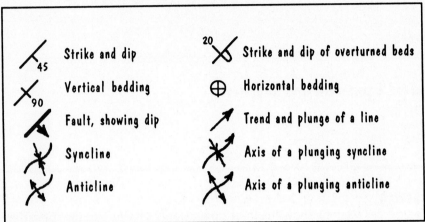

Fig. 2 Geologic map symbols

The orientation of deformed layers can also be obtained through age relationships. Recall that sediments that settle out of water, such as in an ocean or in a bay, are deposited in horizontal layers. The layer at the bottom was deposited first and is therefore the oldest in the sequence of layers. Each new layer is deposited on top of the previous layer. Therefore in a sequence of sedimentary layers the oldest layer is at the bottom of the sequence and the youngest layer is at the top of the sequence. As these sedimentary layers are subjected to stress they fold and tilt. Each fold has an axis. If the tilted layers dip toward the fold axis, the fold is called a **syncline**. The rocks in the center, or core, are younger than those away from the core. If the tilted layers dip away from the axis, the fold is called an **anticline**. The rocks in the core of the fold of an anticline are older than the rocks away from the core (Figure 3).

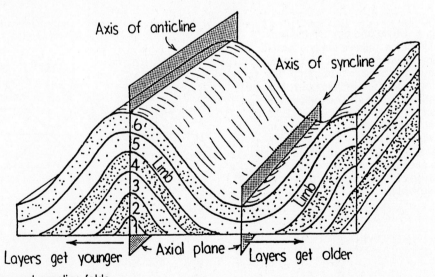

Fig. 3 Anticline and syncline folds.

Over and Under

The fold axis itself can be folded, but more often it is simply tilted. Folds in which the axis is tilted are known as *plunging folds* (Figure 4).

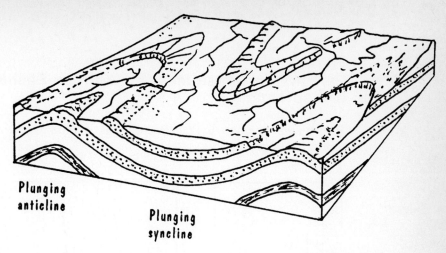

Fig. 4 Plunging folds.

When stress exceeds the mechanical strength of a rock, the rock breaks or faults. The type of fault can be deciphered from the ages of the rocks on either side of the fault. Thrust and reverse faults are produced by compressional (squeezing) forces. These types of faults have older, structurally deeper rocks pushed on top of younger, structurally higher rocks. Normal faults are produced by tensional (pulling) forces. This type of fault has younger, structurally higher rocks on top of older, structurally deeper rocks. The evidence for faulting at the earth's surface is "missing" or "repeated" layers of rock.

Strike-slip faults are produced by shearing forces where the rocks slide past one another. Strike slip faults show no vertical movement, the movement is horizontal. The different types of faults are illustrated in Figure 5.

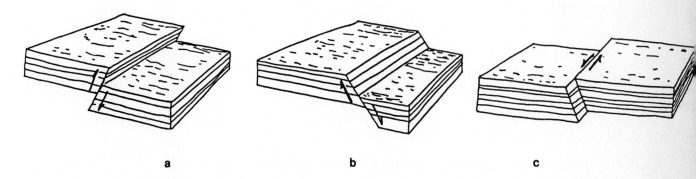

Fig. 5 a Compressive forces produce thrust faults and reverse faults; **b** Tensional forces produce normal faults; **c** Shearing forces produce strike-slip faults.

Age sequence relationships can also be determined from faults and igneous intrusions. When igneous intrusions or faults cut through other rocks, the intrusion or fault must be younger than the rocks they cut.

In this exercise we will use age relationships, strike and dip orientations, and fault evidence to construct cross sections from map views. A map view only shows the surface features — rock type, sequence of rocks, strike and dip, and fault lines. The cross section view is the subsurface extension of the surface view.

Over and Under

179

Procedure
Part A. The Construction of a Cross Section using Strike and Dip Measurements

Step 1: Refer to Figure 6. Transfer locations of contacts and strike and dip symbols to topographic profile.

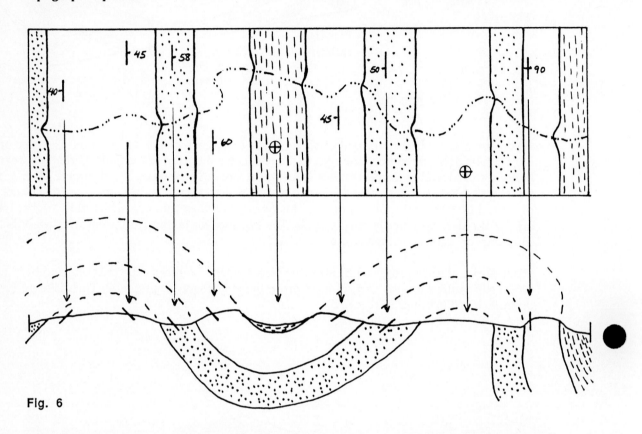

Fig. 6

Step 2: Align the protractor to the location of the transfered dip mark on the topographic profile. Depending on the dip direction, rotate protractor to measured angle. Mark this angle onto profile. Repeat for all dip marks.

Step 3: Using the dip lines on the topographic profile as guides, draw in the bedding contacts. The lines should be smooth and parallel.

Step 4: Dashed lines may be used to show the eroded surface.

Part B. The Construction of Cross Sections using Age Relationships

Step 1: Determine if there is evidence of folding. Folding will show a mirror image about the fold axis of geologic layers.

Step 2: If the beds are folded, determine if the bedding gets younger or older away from the axis of the fold.

Over and Under

Step 3: If the beds are not folded the disturbed sequence may be due to faulting. Are there repeated layers? Are there missing rock layers? Are the layers offset?

Part C. The Construction of Cross Sections Disturbed by Igneous Intrusions

Step 1: Refer to Figure 7. If the beds are not folded, look to see if there has been a disturbance — faulting or igneous intrusion. Faults and intrusions are always younger than the rock they cut into.

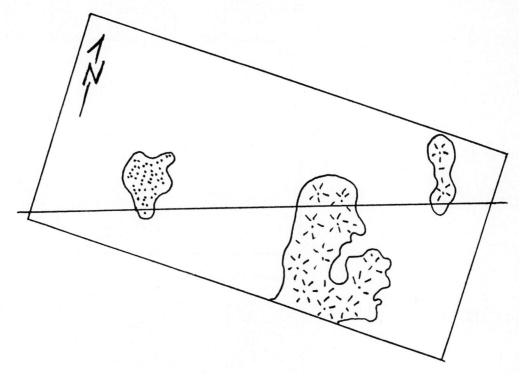

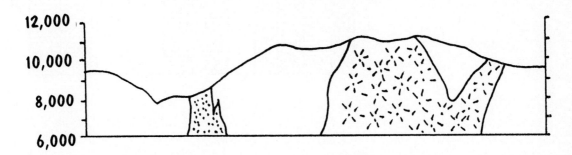

Fig. 7

Step 2: Igneous intrusions work their way up from the deep subsurface. Since there is no way to tell the extent of an intrusion in the subsurface, there is a lot of interpretation.

Over and Under

Exercises

1. Use strike and dip symbols to determine the cross section. What is the structure?

_____

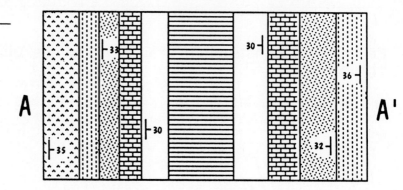

A   A'

2. Use strike and dip symbols to determine the cross section. What type of structure is this?

_____

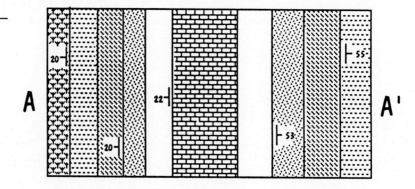

A   A'

Over and Under

3. Use age relationships and general strike and dip symbols to determine the cross section. What type of structure is this?

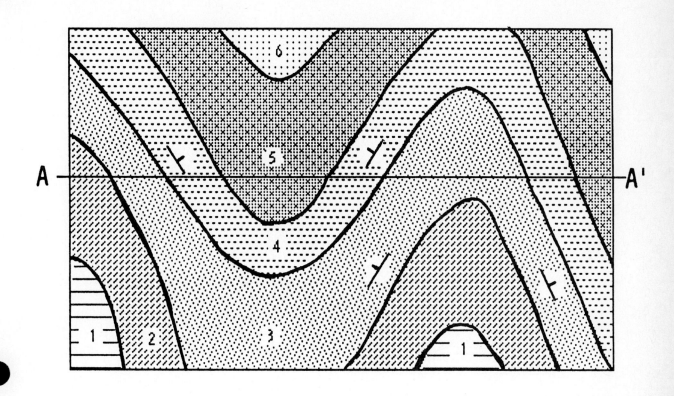

Over and Under

4. Use age relationships to determine cross section. Note that 1 is oldest, 6 is youngest. What type of structure is this?

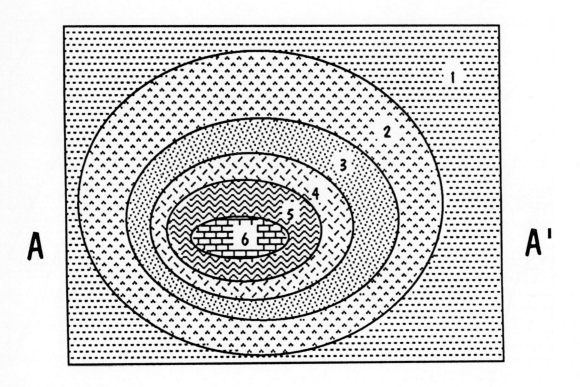

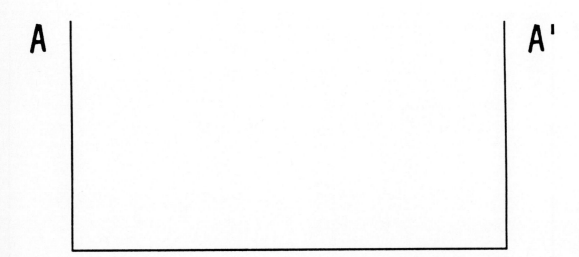

5. Use age relationships to determine cross section. Note that 1 is oldest, 5 is youngest. What type of structure is this?

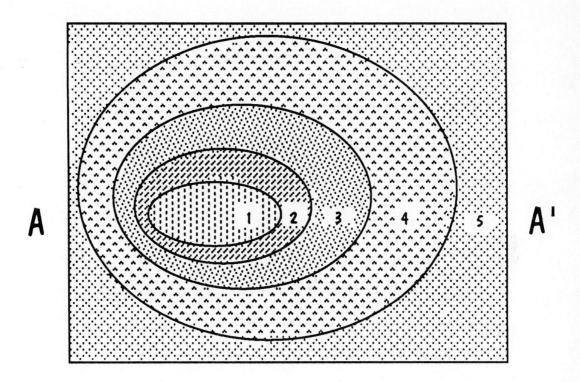

Over and Under

6. This map shows evidence of disturbance and will require some interpretation. Determine the fold structure using the general strike and dip symbols. What type of fold is displayed?

_____

The darkened line represents a fault surface. Show the fault direction movement by placing arrows on the map. What type of fault does this represent?

_____

Draw the cross section. What is the oldest structure? _____

What is the youngest structure?_____

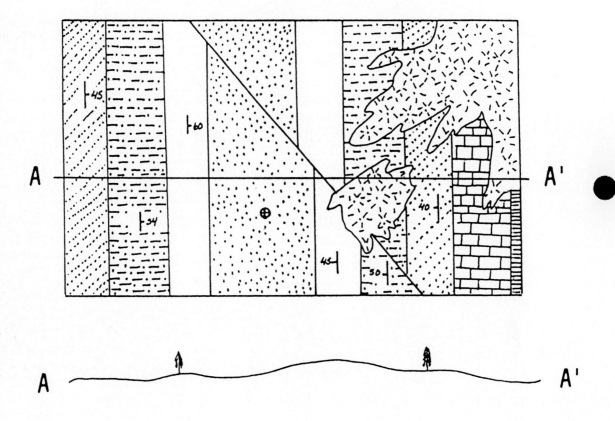

Over and Under

Summing Up

1. What type of structure is displayed in Exercise 1? Is the structure symmetrical or asymmetrical? Can the symmetry be determined from the map view?

2. What type of structure is displayed in Exercise 2? Is the structure symmetrical or asymmetrical? Can the symmetry be determined from the map view?

3. What type of structure is displayed in Exercise 3? What does the dip direction tell us about the structure?

4. Exercise 4 and Exercise 5 are very similar to each other. How are they similar? How are they different? Mark the dip direction for both exercises.

5. What type of fault is displayed in Exercise 6? What evidence supports your answer? What is the fold structure? Based on the drawing, which structure is oldest? Which is youngest?

Over and Under

CONCEPTUAL **Physical Science**

Atmosphere and Ocean Interactions

Solar Power I

Purpose
To measure the sun's power output by comparison with the power output of a 100-watt light bulb.

Required Equipment and Supplies
2-cm by 6-cm piece of aluminum foil with one side thinly coated with flat black paint
clear tape
meter stick
glass jar with a hole in the metal lid
one-hole stopper to fit the hole in the jar lid
fine-scaled thermometer
clear 100-watt light bulb with receptacle

Discussion
To measure the power output of a light bulb is nothing to write
home about — but to measure the power output of the sun using only
household equipment is a different story. In this experiment we'll
do just that — measure to a fair approximation the power output of
the sun. We'll do this by comparing the light from a bulb of known
wattage with the light from the sun, using the simple ratio

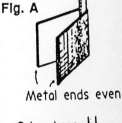
Absorber

$$\frac{\text{sun's wattage}}{\text{sun's distance}^2} = \frac{\text{bulb's wattage}}{\text{bulb's distance}^2}$$

We'll need a sunny day to do this experiment.

Procedure
Step 1: With the blackened side facing out, fold the middle of the foil strip around
the thermometer bulb, as shown in Figure A. The ends of the metal strip should line
up evenly.

Step 2: Crimp the foil so it completely surrounds the bulb, as in Figure B. Bend each
end of the foil strip outward (Figure C). On a tabletop or with a meter stick, make a
flat, even surface. Use a piece of clear tape to hold the foil to the thermometer.

Step 3: Insert the free end of the thermometer into the one hole stopper (soapy
water or glycerin helps). Remove the lid from the jar, and place the stopper in the
lid from the bottom side. Slide the thermometer until the foil strip is located in the
middle of the jar. Place the lid on the jar.

Step 4: Position the jar indoors near a window so that the sun will shine on it. Prop
it at an angle so that the blackened side of the foil strip is perpendicular to the rays
of the sun. Keep it in this position in the sunlight until a stable temperature is
reached. If you prefer, do this outside, which means doing Steps 5 and 6 outside.

Stable temperature = _____ °C

Fig. A

Metal ends even

Crimp here
Fig. B

Fig. C

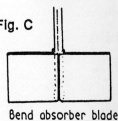
Bend absorber blades
outward

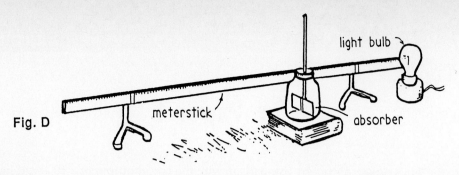

light bulb

Fig. D   meterstick   absorber

Step 5: Now find the conditions for bringing the foil to the same temperature with a clear 100-watt light bulb. Set the meter-stick on the table. Place the clear 100-Watt light bulb with its filament located at the 0-cm mark of the meter-stick (Figure D). Center the jar at the 95-cm mark with the blackened side of the foil strip perpendicular to the light rays from the bulb. You may need to put some books under the jar to align it properly.

Step 6: Turn the light bulb on. Slowly move the jar toward the light bulb, 5 cm at a time, allowing the thermometer temperature to stabilize each time. As the temperature approaches the reading reached in Step 4, move the jar only 1 cm at a time. When the same temperatur obtained from the sun is maintained by the bulb for about two minutes, turn the light bulb oft.

Step 7: Measure as exactly as possible the distance in meters between the foil and the filament of the bulb. Record this distance.

Distance from light filament to foil strip = _____ m

Step 8: Since you know the Sun's distance from the thermometer in meters is 1.5×10^{11} m, and you know the distance of the light bulb from the thermometer, and the wattage of the bulb, you know 3 of the 4 values for the ratio equality stated earlier. With simply rearrangement

$$\text{Sun's wattage} = \frac{(\text{bulb's wattage})(\text{Sun's distance})^2}{(\text{bulb's distance})^2}$$

Show your work:

Sun's wattage = _____ W

Step 9: Use the Sun's wattage to compute the number of 100-watt light bulbs needed to equal the Sun's power. Show your work.

Number of 100 watt light bulbs = _____

Summing Up

1. The accepted value for the Sun's power is 3.8×10^{26} W. How close was your experimental value? List factors you can think of that affect the difference.

_____

_____

_____

CONCEPTUAL **Physical Science** ━━━━━━━━━━━━━━━ **Experiment**

Atmosphere and Ocean Interactions

Solar Power II

Purpose
To measure the amount of solar energy per minute that reaches the earth's surface, and from this estimate the sun's power output.

Required Equipment and Supplies
2 Styrofoam cups
graduated cylinder
water
blue and green food coloring
plastic wrap
rubber band
thermometer (Celsius)
meter stick

Discussion
How do we know how much energy the sun radiates? First we assume that the sun radiates energy equally in all directions. Imagine a heat detector so big that it completely surrounds the sun, like an enormous basketball with the sun at its center. Then the heat energy reaching this detector each second would be the sun's power output. Or if our detector were half a basketball and caught half the sun's energy, then we would multiply the detector reading by two to compute the total solar output. If our detector were a quarter of a basketball, and caught one quarter of the sun's energy, then its total output would be four times the detector reading.

Now that you have the concept, suppose that our detector is the water surface area of a full Styrofoam cup here on earth facing the sun. Then that area is only a tiny fraction of the area surrounding the sun. If we figure what that fraction is, and also measure the energy captured by the water in our cup, we can tell how much total energy the sun radiates. In this experiment, we'll measure the amount of solar energy that reaches a Styrofoam cup and relate it to the amount of solar energy that falls on the earth. We'll need a sunny day for this.

Procedure
Step 1: Using a graduate, measure and record the amount of water needed to fill a Styrofoam cup. Add a small amount of blue and green food coloring to make the water dark (and a better absorber of solar energy) and the cup brimful.

Volume of water = _____ mL

Mass of water = _____ g

Nest the water-filled cup in a second Styrofoam cup (for better insulation).

Step 2: Measure the water temperature and record it.

Initial water temperature = _____ °C

Step 3: Cover the doubled cup with plastic wrap, and seal it with a rubber band to prevent water leaks. Place the cup in the sunlight, tipped if necessary so that the plastic surface is approximately perpendicular to the sun rays. Let it warm by sunlight for 10 minutes.

Step 4: Remove the plastic wrap. Stir the water in the cup gently with the thermometer. Measure the final water temperature and record it. Find the difference in temperature before and after being in the sunlight.

Final water temperature = _____ °C

Temperature difference = _____ °C

THE QUANTITY Q OF HEAT ENERGY COLLECTED BY THE WATER = MASS OF THE WATER × ITS SPECIFIC HEAT ($c = 1 \frac{cal}{g \cdot c}$) × ΔT, ITS CHANGE IN TEMPERATURE

Step 5: Compute the surface area of the top of the cup in square centimeters. (Begin by measuring and recording the diameter of the cup's top in centimeters).

Cup diameter = _____ cm

Surface area of water = _____ cm^2

Step 6: Compute the energy in calories that was collected in the cup ($Q = mc\Delta T$). Assume that the specific heat of the mixture is the same as the specific heat of the water. Show your work on separate paper and record your result here.

Energy = _____ cal

Step 7: Compute the solar energy flux, the energy collected per square centimeter per minute.

Solar energy flux = _____ $\frac{cal}{cm^2 min}$

Step 8: Compute how much solar energy reaches each square meter of the earth per minute. Again, show your work on separate paper. (Hint: There are 10,000 cm^2 in 1 m^2.)

Solar energy flux = _____ $\frac{cal}{m^2 min}$

Step 10: The distance between the earth and the sun is 1.5×10^{11}m, or 1.5×10^{13}cm. The area of a sphere is $4\pi r^2$ so you can calculate the surface area of a "basketball" with radius 1.5×10^{13}cm. And we can divide this area by the surface area of the cup we used in this experiment to find what fraction of the sun's total solar output reached this cup. Then you can calculate the total solar output per minute from your work in Step 7. Do so.

Total solar output per minute = _____ cal

Summing Up

Scientists have measured the amount of solar energy flux just above our atmosphere to be 2 calories per square centimeter per minute (equivalently 1.4 kW/m^2). This energy flux is called the solar constant. Only 1.5 calories per square centimeter per minute reaches the earth's surface after passing through the atmosphere. How did your value compare?

_____

What factors could affect the amount of sunlight reaching the earth's surface, decreasing the solar constant?

_____

Solar Power II

CONCEPTUAL **Physical Science** | Activity |

Weather

Indoor Clouds

Purpose
To illustrate the formation of a cloud from the condensation of water droplets.

Required Equipment and Supplies
large glass jar
measuring cup for water
small metal baking tray
tray of ice cubes

Discussion
Clouds are made up of millions of tiny water droplets and/or ice crystals. Cloud formation takes place as rising moist air expands and cools.

As the sun warms the earth's surface, water is evaporated from the oceans, lakes, streams and rivers. The process of evaporation changes the water molecules from a liquid phase to a vapor phase. Because evaporation is greater over warmer waters than cooler waters, tropical locations have a higher water vapor content than polar locations. Water vapor content is dependent on temperature. For any given temperature there is a limit to the amount of water vapor in the air. This limit is called the *dew point*. When this limit is reached the air is *saturated*. A measure of the amount of water vapor in the air is called *humidity* (the mass of water per volume of air). The amount of water vapor in the air varies with geographic location and may change according to temperature.

Warm air can hold more water vapor before becoming saturated than cooler air. When the air temperature falls, the air becomes saturated — the amount of water vapor the air can hold reaches its limit. As the air cools below the dew point, the water vapor molecules condense onto the nearest available surface. Condensation is the change from a vapor phase to a liquid phase. The condensation of water vapor molecules on small airborne particles produces cloud droplets which, in turn, become clouds .

Most clouds form as air rises, expands, and cools. There are several reasons for the development of clouds. When the temperatures of certain areas of the earth's surface increase more readily than other areas, air may rise from thermal convection. As this warm air rises it mixes with the cooler air above and eventually cools to its saturation point. The moisture from the warm air condenses to form a cloud. As the cloud grows it shades the ground from the sun. This cuts off the surface heating and upward thermal convection — the cloud dissipates. After the cloud is gone the ground once again heats up to start another cycle of thermal convection.

Clouds also form as a result of topography. Air rises as it moves over mountains. As it rises, it cools. If the air is humid, clouds form. As the air moves down the other side of the mountain, it warms. This air is drier since most of the moisture has been removed to form the clouds on the

Indoor Clouds

other side. It is therefore more common to find cloud formation on the windward side of mountains rather than on the leeward side.

Clouds may form as a result of converging air. When cold air moves into a warm air mass, the warm air is forced upward. As it rises, it cools, and water vapor condenses to form clouds. Cold fronts are associated with extensive cloudiness and thunderstorms. When warm air moves into a cold air mass, the less dense warmer air rides up and over the colder denser air. Warm fronts result in widespread cloudiness and light precipitation that may extend for thousands of square kilometers.

Procedure
Step 1:
Fill a glass jar with about one inch of very hot water (do not use boiling water — glass may break).

Step 2:
Place ice cubes in a metal baking tray and set the tray on top of the jar. Make sure there is a good seal.

Step 3:
Observation of a "cloud". As the air inside the top of the jar is cooled by the ice above, water vapor condenses into water droplets to form a "cloud".

Summing Up
1. How is your formation of clouds the same as the formation of clouds in the sky?

_____

_____

2. How is it different?

_____

_____

3. Why are coastal tropical areas more humid than desert areas?

_____

_____

4. Why do we believe that warm air rises?

_____

_____

5. Explain how topography contributes to the formation of desert areas — for example the Sierra Nevada Mountains and the Nevada Desert.

_____

_____

Indoor Clouds

CONCEPTUAL **Physical Science**   **Experiment**

The Solar System

Sunballs

Purpose
To estimate the diameter of the sun.

Required Equipment and Supplies
small piece of cardboard
meterstick

Discussion
Take notice of the round spots of light on the shady ground beneath trees. These are sunballs — images of the sun (discussed in the Prologue of *Conceptual Physical Science*). They are cast by openings between leaves in the trees that act as pinholes. The diameter of a sunball depends on its distance from the small opening that produces it. Large sunballs, several centimeters in diameter or so, are cast by openings that are relatively high above the ground, while small ones are produced by closer "pinholes." The interesting point is that the ratio of the diameter of the sunball to its distance from the pinhole is the same as the ratio of the sun's diameter to its distance from the pinhole.

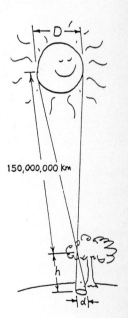

Since the sun is approximately 150,000,000 km from the pinhole, careful measurement of this ratio tells us the diameter of the sun. That's what this experiment is all about. Instead of finding sunballs under the canopy of trees, you'll make your own easier-to-measure sunballs.

Procedure
Poke a small hole in a piece of cardboard with a pen or sharp pencil. Hold the cardboard in the sunlight and note the circular image that is cast on a convenient screen of any kind. This is an image of the sun. Unless you're holding the card too close, note that the solar image size does not depend on the size of the hole in the cardboard (pinhole), but only on its distance from the pinhole to the screen. The greater the distance between the image and the cardboard, the larger the sunball.

150,000,000 km

Position the cardboard so the image exactly covers a dime, or something that can be accurately measured. Carefully measure the distance to the small hole in the cardboard. Record your measurements as a ratio:

$$\frac{\text{diameter of dime}}{\text{distance from dime to pinhole}} = \underline{\hspace{2cm}}$$

Since this is the same ratio as the diameter of the sun to its distance, then

$$\frac{\text{diameter of dime}}{\text{distance from dime to pinhole}} = \frac{\text{diameter of sun}}{\text{distance from sun to pinhole}}$$

Which means you can now calculate the diameter of the sun!

Diameter of the sun = _____

Sunballs

Summing Up

1. Will the sunball still be round if the pinhole is square shaped? Triangle shaped? (Experiment and see!)

_____

_____

_____

2. If the sun is low so the sunball is elliptical, should you measure the small or the long width of the ellipse for the sunball diameter in your calculation of the sun's diameter? Why?

_____

_____

_____

3. If the sun is partially eclipsed, what will be the shape of the sunball?

_____

_____

WHAT SHAPE DO SUNBALLS HAVE DURING A PARTIAL ECLIPSE OF THE SUN?

CONCEPTUAL **Physical Science**

Activity

Planetary Motion

Ellipses

Purpose
To investigate the shapes of ellipses and the locations of their focal points (foci).

Required Equipment and Supplies
about 20 cm of string
2 thumbtacks
pencil and paper
a flat surface that will accept thumbtack punctures

Discussion
An ellipse is an oval-like closed curve defined as the locus of all points about a pair of foci (focal points), the sum of whose distances from both foci is a constant. Planets orbit the sun in elliptical paths, with the sun's center at one focus. The other focus is a point in space, typified by nothing in particular.

Elliptical trajectories are not confined to "outer space." Toss a ball and that parabolic path it seems to trace is actually a small segment of an ellipse. If extended, its path would continue through the earth and swing about the earth's center and return to its starting point. In this case, the far focus is the earth's center. The near focus is not typified by anything in particular. The ellipse is very stretched out, with its long axis considerably longer then its short axis. We say the ellipse is very *eccentric*. Toss the rock faster and the ellipse is wider — less eccentric. The far focus is still the earth's center, and the near focus is nearer to the earth's center than before. Toss the rock at 8 km/s and both foci will coincide at the earth's center. The elliptical path is now a circle — a special case of an ellipse. Toss the rock faster, and it follows an ellipse external to the earth. Now we see the near focus is the center of the earth and the *far* focus is beyond — again, at no particular place. As speed increases and the ellipse becomes more eccentric again, the far focus is *outside* the earth's interior.

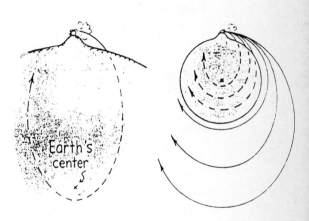

Constructing an ellipse with pencil, paper, string, and tacks, is interesting. Let's do it!

Procedure
Place a loop of string around two tacks or push pins and pull the string taut with a pencil. Then slide the pencil along the string, keeping it taut (Figure 2). (To avoid twisting of the string, it helps if you make the top and bottom half in two separate operations.)

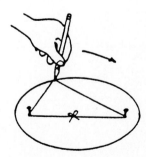

Ellipses

Label each focus of your ellipse (the location of the push pins). Repeat, using different focus separation distances, and you'll notice that the greater the distance between foci, the more eccentric will be the ellipse.

Construct a circle by bringing both foci together. A circle is a special case of an ellipse.

Another special case of an ellipse is a straight line. Determine the positions of the foci that give you a straight line.

The shadow of a ball face on is a circle. But when the shadow is not face on, the shape is an ellipse. The photo below shows a ball illuminated with three light sources. Interestingly enough, the ball meets the table at the focus for all three ellipses.

Summing Up

1. Which of your drawings approximates the earth's orbit about the sun?

_____

2. Which of your drawings approximates the orbit of Haley's comet about the sun?

_____

3. What is your evidence for the definition of the ellipse: that the sum of the distances from the foci is constant?

_____

_____

_____

CONCEPTUAL **Physical Science** ——————————————— **Experiment**

Astronomy Measurements

Reckoning Latitude

Purpose
To build and use the necessary instruments for determining how far above the equator you are on this planet. This will be done by measuring the angle of Polaris in the sky. Actual measurements will need to be made outside of class as measurements can only be performed at nighttime.

Required Material and Supplies
protractor
pencil with unused eraser
pushpin
drinking straw
plumb line and bob
tape

Discussion
Latitude is a measure of your angular distance north or south of the equator (Figure 1). To determine your latitude in the northern hemisphere, it is convenient to use Polaris, the North Star, as a guide. At the North Pole you would find that the North Star is directly above you — 90° above the horizon. This is also your angular distance from the equator (Figure 2a). If you were to travel closer to the equator, the North Star would recede behind you to some lower angle above the horizon, which would always match your angular distance from the equator (Figure 2b). If for example you

Fig. 1 Latitudes are parallel to the equator.

measured the North Star to be 45° above the horizon, you would be at an angular distance of 45° north of the equator. At the equator you'd find the North Star to be 0° above the horizon (Figure 2c). (It turns out atmospheric refraction at this grazing angle complicates viewing.)

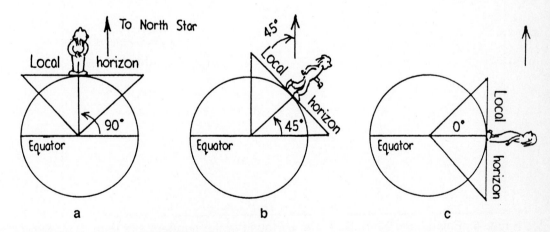

Fig. 2 The angle at which the North Star appears above the horizon is also your angular distance from the equator.

Reckoning Latitude

Procedure

A *theodolite* is a simple instrument for measuring angles above the horizon (altitude). To construct a theodolite, tape a straw to the straight edge of a protractor as shown in Figure 3. Place the eraser end of a pencil against the hole in the protractor and insert the pushpin through the other side of the hole into the eraser — the pencil serves as a convenient handle. Next, tie a plumb bob to the needle (a string attached to a weight).

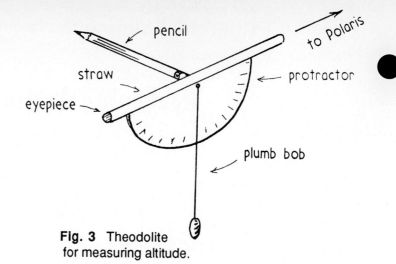

Fig. 3 Theodolite for measuring altitude.

To measure the altitude of any object viewed through the straw, have a partner read the angle the plumb bob line makes with the protractor. Practice using your theodolite by measuring the angle of various objects around you. Work in pairs or small groups so that one person can read the angle while the other is sighting. Take turns looking through the theodolite to be sure of your readings. You may want to make several measurements and find the average. The angle that you read should be between 0° and 90°. Be sure that this angle is not the number of degrees from zenith (straight up). You can convert such readings into degrees from the horizon (straight out) by subtracting the number of degrees from 90°. Challenge yourself or others to estimate measurements without using the theodolite. How accurate are these "naked-eye" estimates?

Measuring the Altitude of Polaris

The North Star is fairly easy to locate because of its position relative to the Big and Little Dippers, two well-known constellations that stand out in the northern sky. To find the North Star from the Big Dipper, draw an imaginary line between the last two stars of the Big Dipper's pan. Follow this line away from the top of the Big Dipper and the first bright star you cross is the North Star (Figure 4). Note that the North Star is the first star of the Little Dipper's handle .

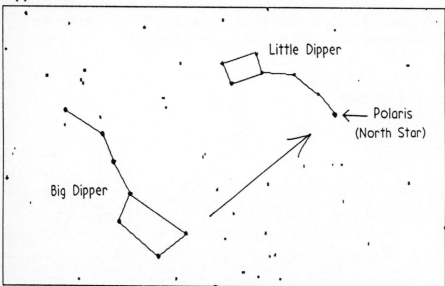

Fig. 4 Polaris, the North Star, can be found in the northern sky from its position relative to the Big and Little Dippers.

Reckoning Latitude

To find your latitude you need simply measure the altitude of the North Star above the horizon. It so happens that the North Star is not *exactly* above the North Celestial Pole, so for a more accurate determination of your latitude, a minor but significant correction may be necessary. The North Star lies about $\frac{3}{4}$ of a degree away from the North Celestial Pole in the direction of Cassiopeia, a "W" shaped constellation (Figure 5). Hence the altitude of the North Star may not correspond exactly to your latitude. If you see Cassiopeia much "above" the North Star (higher up from the horizon), then subtract $\frac{3}{4}$ of a degree from your measured altitude of Polaris to obtain your latitude. If, on the other hand, Cassiopeia is "beneath" the North Star, add $\frac{3}{4}$ of a degree. If Cassiopeia is anywhere to the left or right of the North Star, then the correction is not necessary.

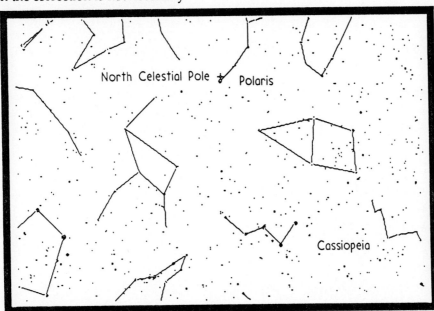

Fig. 5 Polaris is offset from the pole towards Cassiopeia.

Summing Up

1. What was your measured altitude of Polaris?

_____

2. According to this measurement, at what latitude are you located?

_____

3. By how many degrees does your measured latitude differ from the accepted latitude given by your instructor?

_____

4. One degree of latitude is equivalent to 107 km (66 miles). By how many kilometers does your determined latitude differ from the accepted value in kilometers?

(Your Determined Latitude - Accepted Latitude) x 107 = _____

5. How could the theolite be used to provide evidence that the world is round?

_____

Reckoning Latitude

Tracking Mars

Purpose

To plot the orbit of Mars using data obtained by Tycho Brahe four centuries ago.

Required Equipment and Supplies

four sheets of plain grid graph paper
compass
protractor
ruler
sharp pencil

Discussion

In the early 1500s, the Polish astronomer Copernicus used observations and geometry to determine the distance of each planet from the center of the sun and each planet's period of revolution around the sun — on the bold assumption that the planets actually circled the sun. In the late 1500s, before telescopes were invented, the Danish astronomer Tycho Brahe made twenty years of extensive and accurate measurements of planets and bright stars. Near the end of Tycho's career, he hired a young German mathematician, Johannes Kepler, and assigned him the task of plotting Mars' orbit using Tycho's data.

Kepler started by drawing a circle to represent the earth's orbit (not bad, considering the earth's orbit only approximates a circle, which he didn't know at the time). Since Mars takes 687 earth days to orbit the sun once, Kepler paid attention to observations that were exactly 687 days apart. In this way Mars would be in the same place while earth would be in a different location. Two angular readings of Mar's location from the same location on earth 687 days apart was all that was needed — where the two lines crossed was a point on the orbit of Mars. Plotting many such points did not trace out a circle, as Kepler had expected — rather, the path was an ellipse. Kepler was the first to discover that if the planets orbit the sun, they did so in elliptical rather than circular paths. He then went on to plot a better orbit for earth based on observations of the sun, and further refined the plotted orbit of Mars.

In this activity you will duplicate the work of Kepler, simplifying somewhat by assuming a circular orbit for the earth — it turns out the difference is minor, and the elliptical path of Mars is evident. From Brahe's extensive tables, we'll use only data shown in Table A.

Procedure

Step 1: You'll want a sheet of graph paper with about a 14 x 14-inch working area. If need be tape two legal size sheets of borderless graph paper together, or four regular 8 1/2" x 11" sheets.

Step 2: Make a dot at the center of your paper to represent the sun. Place a compass there, and draw a 10-cm radius circle to represent the orbit of the earth around the sun. Draw a light line from the center to the right of the paper, and mark the intersection with the earth's orbit $0°$. This is the position of the earth on March 21st. All your plotting will be counterclockwise around the circle from this reference point.

Table A Tycho's data are grouped in 14 pairs of Mars sightings. For the first 9 pairs, the first line of data is for Mars in *opposition* — when Sun, Earth and Mars were on the same line — when Mars was 90° to the earth horizon, directly overhead at midnight. The second line of data are positions measured 687 days later, when Mars was again in the same place in its orbit, and Earth in a different place, where a different angle was then measured. All angles given in the table read from a 0° reference line — the line from the sun to earth at the vernal equinox, March 21. Mars at points 10-14 are non-opposition sightings. The first line of Point 10, for example, shows that when Earth was at 277°, Mars was seen not directly overhead, but at 208.5° with respect to the 0° reference line. Then 687 days later Earth was at 235°, where Mars was seen at 272.5°. The data are neatly arranged for plotting — something that took Kepler years to do.

Mars Orbit Point	Date Mo	Day	Year	Earth Position (Ecliptic)	Mars Position (Ecliptic)
1	11	28	1580	66.5	66.5
	10	16	1582	22.5	107.0
2	1	7	1583	107.0	107.0
	11	24	1584	62.5	144.0
3	2	10	1585	141.5	141.5
	12	29	1586	97.5	177.0
4	3	16	1587	175.5	175.5
	1	31	1589	132.0	212.0
5	4	24	1589	214.5	214.5
	3	12	1591	171.5	253.5
6	6	18	1591	266.5	266.5
	5	5	1593	225.0	311.0
7	9	5	1593	342.5	342.5
	7	24	1595	300.5	29.5
8	11	10	1595	47.5	47.5
	9	27	1597	4.0	90.0
9	12	24	1597	92.5	92.5
	11	11	1599	48.0	130.5
10	6	29	1589	277.0	208.5
	5	16	1591	235.0	272.5
11	8	1	1591	308.0	260.5
	6	18	1593	266.5	335.0
12	9	9	1591	345.5	273.0
	7	27	1593	304.0	347.5
13	10	3	1593	9.5	337.5
	8	20	1595	327.0	44.5
14	11	23	1593	60.5	350.5
	10	10	1595	17.0	56.0

Step 3: Locate the first point in Mar's orbit, Point 1, from Table A. Do this by first marking with a protractor the positon of the earth along the circle for the date November 28, 1580. This is 66.5° above the 0° reference line. Draw a dot to show earth's position at this time.

Step 4: Mars at this time was in *opposition* — opposite to the sun in the sky. A line from the sun to earth at this time extends radially outward to Mars. Draw a line from the center of your circle (the sun) to the earth's position at this time, and beyond the earth through Mars. Where is Mars along this line? You'll need another sighting of Mars 687 days later when Mars is at the same place and earth is in another.

Step 5: If you were to add 687 days to November 28, 1580, you'd get October 16, 1582. At that date Mars was measured to be lower in the sky — actually 107.0° with respect to the 0° reference line of March 21. With respect to the reference line, use a protractor and ruler and draw a line at 107.0° as shown in Figure A. Where your two lines intersect is a point along Mars' orbit.

Tracking Mars

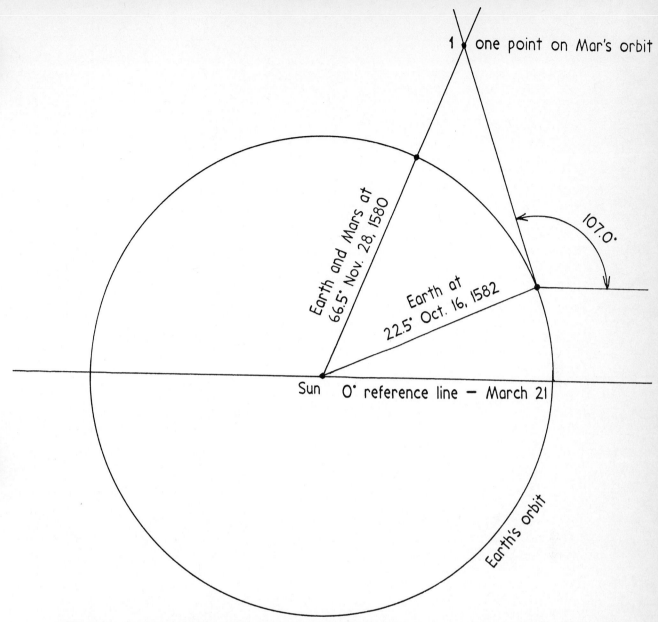

Fig. A Your plot should look like this for finding Mars Orbit Point 1; data are in the first pair of lines of Table A. Angles from earth to Mars are given with respect to the 0° reference line of March 21.

Step 6: With care, plot the 13 other intersections that represent points along Mars' orbit, using the data in Table A.

Step 7: Connect your points, either very carefully by freehand, or with a French curve.

Bravo — you have plotted the orbit of Mars using Kepler's method of four centuries ago!

Summing Up

1. Does your plot agree with Kepler's finding that the orbit is an ellipse? _____

2. During what month are the orbits of Mars and the Earth closest? _____

Tracking Mars

Significant Figures and Uncertainty in Measurement

Units of Measurement

All measurements consist of a number and a unit. Both are necessary. If you say that a friend is going to give you 5, you are telling only *how many*. You also need to tell *what* — five fingers, five cents, five dollars, or five corny jokes. If your instructor asks you to measure the length of a piece of wood, saying that the answer is 26 is not correct. She or he needs to know whether the length is 26 centimeters, feet, or meters. All measurements must be expressed using a number and an appropriate unit.

Numbers

Two kinds of numbers are used in science — those that are counted or defined, and those that are measured. There is a great difference between a counted or defined number and a measured number. The exact value of a counted or defined number can be stated, but the exact value of a measured number cannot.

For example, you can count the number of chairs in your classroom, the number of fingers on your hand, or the number of quarters in your pocket with absolute certainty. Correctly counted numbers are not subject to imprecision or uncertainty.

Defined numbers are about exact relations and are defined to be true. The defined number of centimeters in a meter, the defined number of seconds in an hour, and the defined number of sides on a square are examples. Defined numbers also are also perfectly precise.

Every measured number, no matter how carefully measured, has some degree of uncertainty. What is the width of your desk? Is it 89.5 centimeters, 89.52 centimeters, 89.520 centimeters or 89.5201 centimeters? You cannot state its exact measurement with absolute certainty.

Uncertainty in Measurement

Uncertainty (or margin of error) in a measurement can be illustrated by the two different metersticks in Figure A. The measurements are of the length of a table top. Assuming that the zero end of the meterstick has been carefully and accurately positioned at the left end of the table, how long is the table?

Figure A

The upper scale in the figure is marked off in centimeter intervals. Using this scale you can say with certainty that the length is between 51 and 52 centimeters. You can say further that it is closer to 51 centimeters than to 52 centimeters; you can estimate it to be 51.2 centimeters.

The lower scale has more subdivisions and has a greater precision because it is marked off in millimeters. With this meterstick you can say that the length is definitely between 51.2 and 51.3 centimeters, and you can estimate it to be 51.25 centimeters.

Note how both readings contain some digits that are known, and one digit (the last one) that is estimated. Note also that the uncertainty in the reading of the lower meterstick is less than that of the top meterstick. The lower meterstick can give a reading to the hundredths place, and the top meterstick to the tenths place. The lower meterstick is more *precise* than the top one. So, digits tell us the magnitude of a measurement while the location of the decimal point tells us the precision.

Significant Figures

Significant figures are the digits in any measurement that are known with certainty plus the one digit that is estimated and hence is uncertain. The measurement 51.2 centimeters (made with the top meterstick in Figure A) has three significant figures, and the measurement 51.25 centimeters (made with the lower meterstick) has four significant figures. The right-most digit is always an estimated digit. Only one estimated digit is ever recorded as part of a measurement. It would be incorrect to report the length of the table (Figure A) is 51.253 centimeters as measured with the lower meterstick. This five-significant-figure value would have two estimated digits (the 5 and 3) and would be incorrect because it indicates a *precision* greater than the meterstick can obtain.

Standard rules have been developed for writing and using significant figures, both in measurements and in values calculated from measurements.

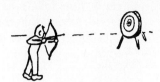

Rule 1
In numbers that do not contain zeros, all the digits are significant.

> EXAMPLES:
> | | |
> |---|---|
> | 4.1327 | five significant figures |
> | 5.14 | three significant figures |
> | 369 | three significant figures |

Rules 2
All zeros between significant digits are significant.

> EXAMPLES:
> | | |
> |---|---|
> | 8.052 | four significant figures |
> | 7059 | four significant figures |
> | 306 | three significant figures |

GOOD PRECISION
BUT
POOR ACCURACY

Rule 3
Zeros to the left of the first nonzero digit serve only to fix the position of the decimal point and are not significant.

> EXAMPLES:
> | | |
> |---|---|
> | 0.0068 | two significant figures |
> | 0.0427 | three significant figures |
> | 0.0003506 | four significant figures |

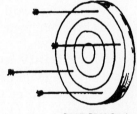

POOR PRECISION
AND
POOR ACCURACY

Rule 4
In a number with digits to the right of the decimal point, zeros to the right of the last non-zero digit are significant.

> EXAMPLES:
> | | |
> |---|---|
> | 53 | two significant figures |
> | 53.0 | three significant figures |
> | 53.00 | four significant figures |
> | 0.00200 | three significant figures |
> | 0.70050 | five significant figures |

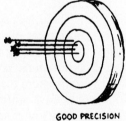

GOOD PRECISION
AND
GOOD ACCURACY

Rule 5
In a number that has no decimal point and that ends in one or more zeros (such as 3600), the zeros that end the number may or may not be significant.

The number is ambiguous in terms of significant figures. Before the number of significant figures can be specified, further information is need about how the number was obtained. If it is a measured number, the zeros are not significant. If the number is a defined or counted number, all the digits are significant.

Confusion is avoided when numbers are expressed in scientific notation. All digits are taken to be significant when expressed this way.

Appendix I

EXAMPLES:

4.6×10^{-5}	two significant figures
4.60×10^{-5}	three significant figures
4.600×10^{-5}	four significant figures
2×10^{-5}	one significant figures
3.0×10^{-5}	two significant figures
4.00×10^{-5}	three significant figures

Rounding off

Calculators often display eight or more digits. How do you round off such a display to, say three significant figures? Three rules govern the process of deleting unwanted (nonsignificant) digits from a calculator number.

Rule 1
If the first digit to the right of the last significant figure is less than 5, that digit and all the digits that follow it are simply dropped.

EXAMPLE:

51.234 rounded off to three significant figures become 51.2.

Rule 2
If the first digit to be dropped is a digit greater than 5, or if it is a 5 followed by a digit other than zero, the excess digits are dropped and the last retained digit is increased in value by one unit.

EXAMPLE:

51.35, 51.359, and 51.3598 rounded off to three significant figures all become 51.4.

Rule 3
If the first digit to be dropped is a 5 not followed by any other digit, or if it is a 5 followed only by zeros, an odd-even rule is applied.

That is, if the last retained digit is even, its value is not changed, and the 5 and any zeros that follow are dropped. But if the last digit is odd, its value is increased by one. The intention of this odd-even rule is to average the effects of rounding off.

EXAMPLES:

74.2500 to three significant figures becomes 7.2
89.3500 to three significant figures becomes 89.4.

Significant Figures and Calculated Quantities

Suppose that you measure the mass of a small wooden block to be 2 grams on a balance, and you find that its volume is 3 cubic centimeters by dipping it beneath the surface of water in a graduated cylinder. The density of the piece of wood is its mass divided by its volume. If you divide 2 by 3 on your calculator, the reading on the display is 0.6666666. However, it would be incorrect to report that the density of the block of wood is 0.6666666 gram per cubic centimeter. To do so would be claiming a higher degree of precision than is warranted. Your answer should be rounded off to a sensible number of significant figures.

The number of significant figures allowable in a calculated result depends on the number of significant figures in the measured data, and on the type of mathematical operation(s) used in calculating. There are separate rules for multiplication and division, and for addition and subtraction.

Appendix I

Multiplication and Division

For multiplication and division an answer must have the number of significant figures found in the number with the fewest significant figures. For the density example given above, the answer must be rounded off to one significant figure, 0.7 gram per cubic centimeter. If the mass were measured to be 2.0 grams, and if the volume were still taken to be 3 cubic centimeters, then the answer must still be rounded off to 0.7 gram per cubic centimeter. If the mass were measured to be 2.0 and the volume 3.0 or 3.00 cubic centimeters, the answer must be rounded off to two significant figures: 0.67 gram per cubic centimeter.

Study the following examples. Assume that the numbers being multiplied or divided are measured numbers.

EXAMPLE A:

8.536 × 0.47 = 4.01192 (calculator answer)

In input with the fewest significant figures is 0.47, which has two significant figures. Therefore, the calculator answer 4.01192 must be rounded off to 4.0.

EXAMPLE B:

3840 ÷ 285.3 = 13.45916 (calculator answer)

The input with the fewest significant figures is 3940, which has three significant figures. Therefore, the calculator answer 13.45916 must be rounded off to 13.5.

EXAMPLE C:

36.00 ÷ 3.000 = 12 (calculator answer)

Both inputs contain four significant figures. Therefore, the correct answer must also contain four significant figures, and the calculator answer 12 must be written as 12.00. In this case the calculator gave too few significant figures.

Addition and Subtraction

For addition and subtraction the answer should not have digits beyond the last digit position common to all the numbers being added or subtracted. Study the following examples:

EXAMPLE A:

$$
\begin{array}{r}
34.6 \\
18.8 \\
+ \ \underline{15} \\
68.4 \text{ (calculator answer)}
\end{array}
$$

The last digit position common to all numbers is that just to the left of where a decimal point is placed or might be placed. Therefore, the calculator answer of 68.4 must be rounded off to 68.

EXAMPLE B:

$$
\begin{array}{r}
20.02 \\
20.002 \\
+ \ \underline{20.0002} \\
60.0222 \text{ (calculator answer)}
\end{array}
$$

The last digit position common to all numbers is the hundredths place. Therefore, the calculator answer of 60.0222 must be rounded off to 60.02.

Appendix I

EXAMPLE C:

345.56
- 245.5
100.06 (calculator answer)

The last digit position common to both numbers in this subtraction is the tenths place. Therefore, the answer must be rounded off to 100.1.

Percentage Error

A measured value is best compared to an accepted value by the percentage of difference rather than the size of the difference. Measuring the length of a 10-centimeter pencil to +/- one centimeter is quite a bit different from measuring the length of a 100-meter track to the same +/- centimeter. The measurement of the pencil shows a relative uncertainty of 10%. The track measurement is uncertain by only 1 part in 10,000, or 0.01%.

The relative uncertainty or relative margin of error in measurements, when expressed as a percentage, is the *percentage of error*. It tells by what percentage a quantity differs from a known accepted value as determined by skilled observers using high precision equipment. It is a measure of the *accuracy* of the method of measurement, which includes the skill of the person making the measurement. The percentage of error is equal to the difference between the measured value and the accepted value of a quantity divided by the accepted value, and then multiplied by 100. (Note: % error should always be presented as a positive number. Take the absolute value if necessary.)

$$\% \text{ error } = \frac{(\text{accepted value} - \text{measured value})}{(\text{accepted value})} \times 100$$

For example, suppose that the measured value of the acceleration of gravity is found to be 9.44 m/s^2. The accepted value is 9.81 m/s^2. The percentage error is

$$\% \text{ error } = \frac{(9.81 \text{ m/s}^2 - 9.44 \text{ m/s}^2)}{(9.81 \text{ m/s}^2)} \times 100 = 3.77\%$$

Appendix I

211

Graphing

Many quantities are dependent on one another. The circumference of a circle, for example, depends on the circle's diameter, and vice versa. Tables, equations, and graphs show how dependent quantities are related. Investigating relationships between dependent quantities make up much of the work of physical science. Tables, equations, and graphs are important physical science tools.

Tables give values of dependent variables in list form. Table A, for example, lists the instantaneous speed of an object relative to its elapsed time of motion. A table organizes experimental data and helps the investigator to deduce relationships.

Table A

Elapsed Time (seconds)	Instantaneous Speed (meters/second)
0	0
1	10
2	20
3	30
4	40
.	.
.	.
.	.
t	$10t$

The relationship between two or more dependent variables can be described using words or can be more concisely expressed using an *equation*. For example, to say that the speed of free fall depends on acceleration and time is said concisely by the equation $v = v_0 + gt$, where v_0 is the initial speed when time $t = 0$. When free fall begins from $v_0 = 0$ we have just $v = gt$.

A *graph* is a pictorial representation of the relationship between dependent variables, such as those found in an equation. By looking at the shape of a graph, we can quickly tell a lot about how the variables are related. For this reason, graphs can help clarify the meaning of an equation or table of numbers. Also, a graph can help reveal the relationship between variables when the equation is not already known. Experimental data is often graphed for this reason.

The most common and simplest graph is the Cartesian graph, where values of one variable are represented on the vertical axis, called the y axis, and values of the other variable are represented on the horizontal or x axis. If variable y is *directly proportional* to variable x then the curve will appear to rise from left to right (Figure A). Immediately you can tell that as x increases, y also increases. If, however, x is *inversely proportional* to y then the curve will appear to lower from left to right (Figure B). In this case, as x increases, y decreases.

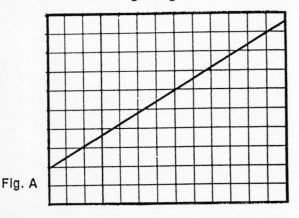

Fig. A   Fig. B

Direct and inverse proportionalities are types of linear relationships. Linear relationships have straight-line graphs — the easiest kind to interpret. Figure C shows a graph of the equation $v = gt$. Speed v is plotted along the y-axis, and time t along the x-axis. As you can see, there is a linear relationship between v and t. Here v increases in direct proportion to t; double t and v doubles, triple t and v triples, and so on.

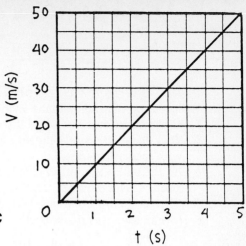

Fig. C

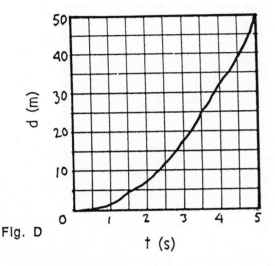

Fig. D

Figure D shows a graph of distance d compared to time t in the equation $d = 1/2 \, gt^2$. We see by the non-linear curve that d is not directly proportional to t. Interestingly enough, if we make a graph of d versus t^2, a straight line results. This is because distance d *is* directly proportional to t^2, with a slope that is calculated to be 5 m/s^2, or $1/2 \, g$ (try it yourself and see). Plotting various powers of non-linear quantities until they form a straight line is one way to find the equation that relates the quantities.

Area Under the Curve

An important feature of a graph that often has physical significance is the area under the curve. Consider the rectangular area bounded by the graph in Figure E. Here the area is the product of the y and x dimensions, speed v and time t respectively (the area of any rectangle is the product of two perpendicular sides). In this case the area vt has physical significance, for $vt = d$, the distance traveled during the time interval t. 50 m/s x 5 s = 250 m, so we see the distance traveled in 5 s is 250 m.

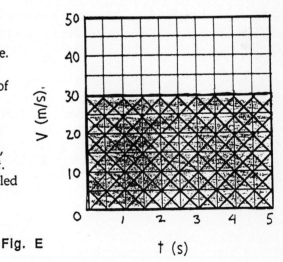

Fig. E

The area need not be rectangular. The area beneath any curve of v versus t represents the distance traveled. Similarly, the area beneath a curve of acceleration versus time gives the velocity acquired ($\Delta v = at$), or the area beneath a force versus time curve gives the momentum acquired ($Ft = \Delta mv$). [What does the area beneath a force versus distance curve give?] The area under various curves, including rather complicated ones, can be found by way of an important branch of mathematics — *integral calculus*.

Appendix II

Graphing with Conceptual Physical Science

Graphs clarify equations by making them pictorial. You will develop some rudimentary graphing skills in your physical science laboratory. Of particular worth is the lab activity that utilizes a ranging device and a computer, *Graphing with Sonar*.

Questions* Figure F is a graphical representation of a ball dropped down the side of a cliff.
1. How long did the ball take to hit the bottom?
2. What was its velocity when it struck bottom?
3. What does the decreasing slope of the graph tell you about the acceleration of the ball with increasing speed?
4. Did the ball reach terminal velocity before hitting the bottom? If so, about how many seconds were required for it to reach its terminal veocity?
5. What is the approximate height of the cliff?
6. How sudden was the impact?

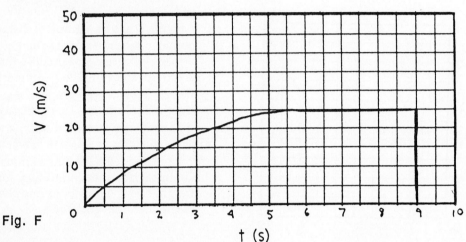

Fig. F

Plotting a Graph

The first step in plotting experimental data on a graph is to choose the dimension and scale of the *x* and *y* axes. The larger a graph the better its precision. For this reason, the axes should be drawn and labeled so that the data is spaced out over as much of the graph paper as possible. The axes should also be labeled with the units they represent.

Data from a data table is entered into the graph as a series of points. Each point represents the magnitudes — as specified by the axes — of both variable quantities (Figure G). If there is a relation between the variables, then as the data is being plotted, a pattern emerges. In Figure G we see the pattern is a straight line.

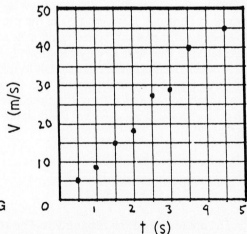

Fig. G

---

*Answers: 1) 9 s; 2) 25 m/s; 3) Acceleration decreases as velocity increases (because of air resistance); 4) Yes (because slope curves to zero), about 6 s; 5) Falling distance is nearly 180 m [the area under the curve is about 71 squares — each square represents 2.5 m (5 m/s x 0.5 s = 2.5 m)]; 6) The impact was very sudden as evidenced by the rapid decrease in velocity at the 9th second (what would the graph look like if the ball bounced back up?).

Appendix II

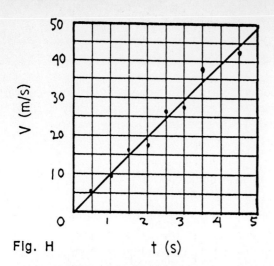

Fig. H

Due to experimental error, which includes the limited precision of instruments, data points often deviate from the apparent pattern. For example, data points showing the pattern of a straight line may not fall exactly in a straight-line. In this case a single straight line that approximates all of the data points may be drawn using a ruler or straight-edge (Figure H). This straight line need not touch any of the data points for it represents the best fit of all the points together. Non-linear relationships can similarly be sketched with a curve that best fits all the data points.

The *slope* of a linear relationship is found by measuring how high the line on the graph rises compared to how far out it extends. This may be done by drawing a right-triangle that has the sloping line as its hypotenuse (Figure I). The slope is equal to the height of the triangle (its "rise") divided by the base (its "run").
Slope = rise/run.

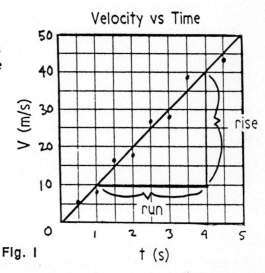

Fig. I

Rise is read as the height the right-triangle rises relative to the y-axis and run is read as the distance the triangle runs relative to the x-axis. The units of the x and y axes must be included. In Figure I, for example, the rise equals 30 m/s and the run equals 3 s. The slope of the line, therefore, equals:

$$slope = \frac{rise}{run} = \frac{30\,m/s}{3s} = 10\frac{m/s}{s} = 10\frac{m}{s^2}$$

The units show us what the slope represents. In this case the sloped line represents the acceleration (m/s²) of the object. Note the distinction between the *angle* of the line, 45°, and the *slope* of the line, 10 m/s².

The final step to creating a graph is assigning a brief but descriptive title, such as "Velocity -vs-Time" (Figure I).

Appendix II

215

Conversion Factors

In the physical sciences it is often necessary to convert one unit into another. Units are converted by mulitplying the given quantity by a *conversion factor*. A conversion factor is a ratio derived from an equality, hence, a conversion factor always equals 1. Consider the following example:

$$12 \text{ inches} = 1 \text{ foot}$$

A given quantity divided by the same quantity always equals one.

$$\frac{12 \text{ inches}}{12 \text{ inches}} = 1 \qquad \text{or} \qquad \frac{1 \text{ foot}}{1 \text{ foot}} = 1$$

Since 12 inches and 1 foot represent the same quantity we can perform the following substitutions:

$$\frac{12 \text{ inches}}{1 \text{ foot}} = 1 \qquad \text{or} \qquad \frac{1 \text{ foot}}{12 \text{ inches}} = 1$$

The above two ratios are conversion factors. Since conversion factors are equal to 1, multiplying a quantity by a conversion factor does not change that quantity. What does change are the units. For example, suppose some item was measured at 60 inches in length. This quantity may be converted into feet by mulitplying by the appropriate conversion factor:

$$(60 \text{ inches}) \; \frac{(1 \text{ foot})}{(12 \text{ inches})} \; = \; 5 \text{ feet}$$

| quantity | conversion | quantity |
| in inches | factor | in feet |

Note how the units of inches cancel each other since they are found in both the numerator and denominator. Similarly, the length of an item measured in feet can be converted to inches using the reciprocal conversion factor.

$$(5 \text{ feet}) \; \frac{(12 \text{ inches})}{(1 \text{ foot})} \; = \; 60 \text{ inches}$$

| quantity | conversion | quantity |
| in feet | factor | in inches |

To derive a conversion factor consult a table of unit equalities, such as Table A. The unit you want to obtain in your answer should be set in the numerator of the conversion factor, and the unit you want to cancel should be set in the denominator.

Several conversion factors may be linked together when the relationship between two units is not given directly. For example, suppose you are to convert 5.00 inches into kilometers. In this case you might convert inches into centimeters and then centimeters into meters followed by meters into kilometers.

$$(5.00 \text{ inches}) \; \frac{(2.54 \text{ cm})}{(1 \text{ inch})} \; \frac{(1 \text{ meter})}{(100 \text{ cm})} \; \frac{(1 \text{ km})}{(1000 \text{ meters})} \; = 0.000127 \text{ kilometers}$$

Alternatively, you might convert inches into feet and then feet into miles followed by miles into kilometers.

$$(5.00 \text{ inches}) \; \frac{(1 \text{ foot})}{(12 \text{ inch})} \; \frac{(1 \text{ mile})}{(5280 \text{ feet})} \; \frac{(1.609 \text{ km})}{(1 \text{ mile})} \; = 0.000127 \text{ kilometers}$$

The technique of following your units guides you in selecting the proper conversion factor. For example, suppose we wish to convert 15.0 calories per minute into joules per second, which are the units of watts. Then

$$(15.0 \frac{cal}{min}) \frac{(4.187 \ J)}{(1 \ cal)} \frac{(1 \ min)}{(60 \ sec)} = 1.22 \frac{J}{sec} = 1.22 \ watts$$

Similarly, we may convert 55.0 miles per hour into meters per second:

$$(55.0 \frac{mi}{hr}) \frac{(5280 \ ft)}{(1 \ mi)} \frac{(12 \ in)}{(1 \ ft)} \frac{(2.54 \ cm)}{(1 \ in)} \frac{(1 \ m)}{(100 \ cm)} \frac{(1 \ hr)}{(3600 \ sec)} = 24.6 \frac{m}{sec}$$

In all cases, non-wanted units cancel and only wanted units remain.

Table A   Units of Measurement

UNITS OF LENGTH		EXACT ?*
1 kilometer (km)	= 1000 meters (m)	yes
1 meter (m)	= 1000 millimeters (mm)	yes
	= 100 centimeters (cm)	yes
1 micron (u)	= 1×10^{-6} m	yes
1 nanometer (nm)	= 1×10^{-9} m	yes
1 angstrom (A)	= 1×10^{-10} m	yes
1 inch (in)	= 2.54 cm	yes
1 mile	= 1.6093440 km	yes
	= 5280 feet	yes
UNITS OF MASS AND WEIGHT		
1 kilogram (kg)	= 1000 g	yes
1 gram (g)	= 1000 milligrams (mg)	yes
1 kg	= 2.205 pounds	no
1 pound (lb)	= 453.6 g	no
	= 16 ounces (oz)	yes
	= 4.448 newtons	no
UNITS OF VOLUME		
1 liter (L)	= 1000 milliliters (mL)	yes
1 mL	= 1 cubic centimeter (cc or cm^3)	yes
1 L	= 1.057 quarts	no
1 cubic inch (in^3)	= 16.39 mL	no
1 gallon (gal)	= 4 quarts (qt) = 8 pints (pt)	yes
UNITS OF ENERGY		
1 calorie (cal)	= 4.184 joules (J)	yes
	= amount of heat needed to raise 1 g of water by 1 °C	yes
1 kilocalorie (kcal)	= 1000 cal = 1 nutritional calorie (C)	yes
	= 4.184 kJ	yes
1 joule	= 1×10^7 erg = 1 kg m²/s²	yes
1 electron volt (ev)	= 24.217 kcal/mol = 1.602×10^{-19} J	no
1 kilowatt hour (kWh)	= 3.60×10^6 J	no
UNITS OF PRESSURE		
1 atmosphere (atm)	= 760 mm Hg = 760 torr	yes
	= 14.70 pounds per square inch (psi)	no
	= 29.29 in. Hg	no

*Conversion factors involving exact relationships contain an unlimited number of significant figures (See appendix I).

Exercises:
1. Convert 20.0 calories into joules

(20.0 calories)

    ⇑   ⇑   ⇑

  quantity   conversion   quantity
  in calories   factor   in joules

2. Convert 26.2 miles into kilometers

(26.2 miles)

    ⇑   ⇑   ⇑

  quantity   conversion   quantity
  in miles   factor   in km

3. Convert 2.00 miles into centimeters

- - - - - - - - - - - - - - - -

Answers:
1. Use the conversion factor 4.187 joules/ 1 calorie to convert 20.0 calories into 83.7 joules.

2. Use the conversion factor 1.609 km/1 mile to convert 26.2 miles into 42.2 kilometers.

3. Convert miles into feet and then feet into inches followed by inches into centimeters to get the answer 322,000 centimeters (See appendix I for rules concerning significant figures).

Appendix III